FRONTIERS IN EARTH SCIENCES

Subduction Zone Magmatism

FRONTIERS IN EARTH SCIENCES

Forthcoming titles in the series:

N. Christie-Blick and M. Kominz: Concepts and Techniques in Physical Stratigraphy

K. D. Collerson and S. A. Bowring: Early Evolution of the Earth

P.G. DeCelles and G. Mitra: Thrust Belts and Synorogenic Sediments

K. Hodges: Metamorphic Pressure—Temperature—Time Paths

FRONTIERS IN EARTH SCIENCES

Subduction Zone Magmatism

Yoshiyuki Tatsumi, PhD
School of Earth Sciences
Faculty of Integrated Human Studies
Kyoto University
Japan

Steve Eggins, PhD
Research School of Earth Sciences
The Australian National University
Institute of Advanced Studies
Canberra, Australia

Blackwell Science

Blackwell Science

EDITORIAL OFFICES:
238 Main Street, Cambridge, Massachusetts 02142, USA
Osney Mead, Oxford OX2 0EL, England
25 John Street, London WC1N 2BL, England
23 Ainslie Place, Edinburgh EH3 6AJ, Scotland
54 University Street, Carlton, Victoria 3053, Australia
Arnette Blackwell SA, 1 rue de Lille, 75007 Paris, France
Blackwell Wissenschafts-Verlag GmbH, Kurfürstendamm 57, 10707 Berlin, Germany
Blackwell MZV, Feldgasse 13, A-1238 Vienna, Austria

DISTRIBUTORS:

USA
Blackwell Science, Inc.
238 Main Street
Cambridge, Massachusetts 02142
(Telephone orders: 800-215-1000 or 617-876-7000)

CANADA
Oxford University Press
70 Wynford Drive
Don Mills, Ontario
M3C 1J9
(Telephone orders: 416-441-2941)

AUSTRALIA
Blackwell Science Pty Ltd
54 University Street
Carlton, Victoria 3053
(Telephone orders: 03-347-5552)

OUTSIDE NORTH AMERICA AND AUSTRALIA
Blackwell Science, Ltd.
c/o Marston Book Services, Ltd.
P.O. Box 87
Oxford OX2 0DT
England
(Telephone orders: 44-865-791155)

Acquisitions: Jane Humphreys
Development: Debra Lance
Production: Michelle Choate
Manufacturing: Kathleen Grimes
Designed by Joyce C. Weston
Typeset by Best-set Typesetter Ltd., Hong Kong
Printed and bound by Braun-Brumfield, Inc., Ann Arbor, MI

Printed in the United States of America

95 96 97 98 5 4 3 2 1

Library of Congress Cataloging in Publication Data

Tatsumi, Yoshiyuki.
Subduction zone magmatism / Yoshiyuki Tatsumi, Steve Eggins.
p. cm.—(Frontiers in earth sciences)
Includes bibliographical references and index.
ISBN 0-86542-361-X
1. Magmatism. 2. Subduction zones. I. Eggins, Steve. II. Title. III. Series.
QE461.T273 1995
552′. 1—dc20 94-44612
CIP

Contents

Preface

Subduction zones are major sites of volcanism on the Earth. They are located at the junction of two converging tectonic plates, where one sinks (subducts) beneath the other into the Earth's mantle. The flux of magma to the overriding plate provides a significant contribution to the growth of the continental crust in the modern Earth, and possibly has done so through much of Earth's history. Subduction of the lower plate is fundamental to both the thermal evolution and chemical differentiation of the mantle and to the Earth itself. It is now clear that subduction zones need to be considered not only in their shallow upper mantle and crust context but on the scale of the whole mantle, as evidence accumulates for subducting plates extending into the lower mantle and probably reaching the base of the mantle. To understand the dynamic evolution of the Earth, we need to understand the physical and chemical consequences of the "entire" subduction process.

The interaction of physical and chemical processes during subduction is complex. Faced with this, our approach has been to focus on what we believe to be the dominant processes which determine the first-order characteristics of the system. Our understanding has developed through the familiar cycle of model formulation, prediction, followed by comparison of observation and prediction, then model refinement. Our current preferred model remains simple, and in this regard we appreciate its shortcoming in the presence of natural complexity. The book is structured about the model, so it neither covers all possible interpretations nor encompasses the complexities of data or data types now available. It inherits our biases, though we have tried to alert readers to valid alternatives. Some specialists may find our approach limited, or that it does not explain observations in a particular area. Nonetheless we hope that it provides a foundation of current knowledge for advance through critique and refinement.

This book is dedicated to the late Masanori Sakuyama for his insights, his emphasis on the importance of modeling basic observations, and for leading the first author to such an interesting field of earth science.

We thank Yukako Higuchi for preparing the manuscript and figures, and Yoshitsugu Furukawa for constructive discussions on mantle dynamics, and we need to thank Keiko and Tricia for their patience.

Yoshiyuki Tatsumi
Kyoto
Steve Eggins
Canberra
December 1993

FRONTIERS IN EARTH SCIENCES

Subduction Zone Magmatism

1

Subduction Zone Magmatism

1.1 Introduction

Subduction zones are regions of the Earth affected by the sinking (subduction) of relatively cold and dense lithospheric plates into the Earth's mantle. The subduction process is commonly manifested as magmatism and surface volcanism (Figure 1.1), such as that which rings much of the Pacific Ocean and occurs along the northeast margin of the Indian Ocean. The global distribution of subduction zones is shown in Figure 1.2.

Subduction is of primary geodynamical and geochemical significance to the modern Earth. Injection of cold lithosphere into the mantle cools the Earth's interior and, accordingly, is a key component of mantle convection. Subduction also introduces geochemical heterogeneity into the mantle through the incorporation of basaltic oceanic crust, geochemically depleted mantle peridotite, and seafloor sediments. Subduction zone magmatism is itself a major contributor to the growth of new continental crust and, thus, to gross chemical differentiation of the Earth's mantle/crust system. Interest in subduction zones and subduction zone magmatism is not limited to these large-scale processes, but extends to microscale levels and includes pursuits as diverse as unraveling associations between certain ore deposit and magma types and the hazard potential posed by the often violent explosive eruptions which characterize many subduction zone volcanoes.

In this book we focus on the origin of subduction zone magmas. We begin by considering several simple, key observations that we believe provide fundamental insights into the unseen processes

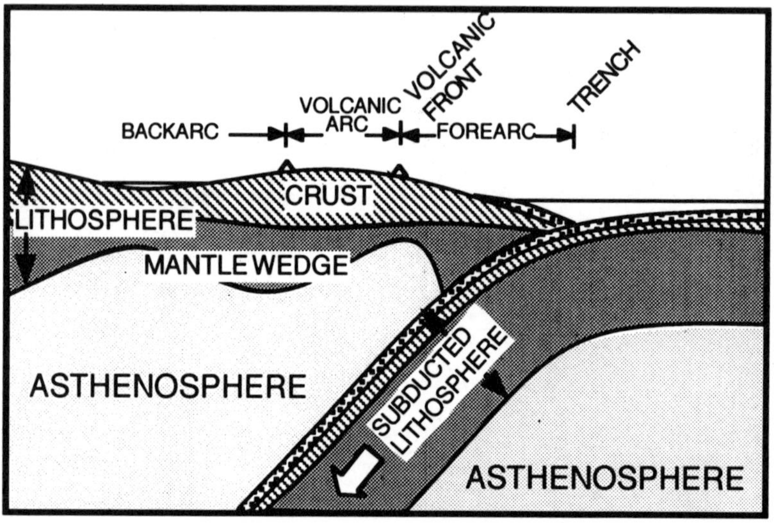

Figure 1.1 A schematic cross section through a subduction zone, illustrating the three regional divisions; the forearc, volcanic arc, and backarc. The volcanic front, the line of volcanoes nearest the trench, forms the boundary between the volcanic arc and forearc.

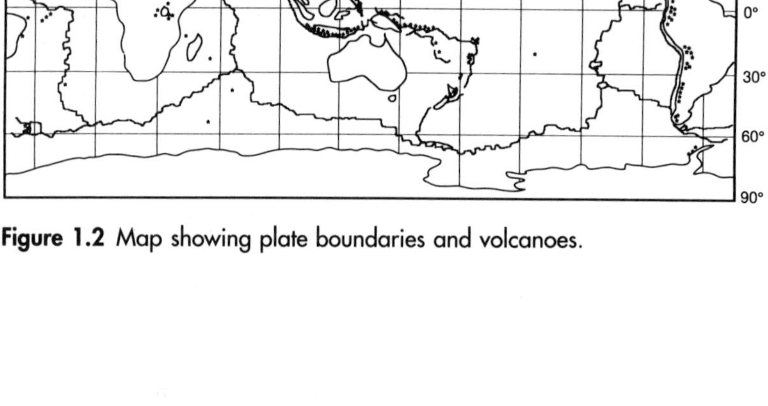

Figure 1.2 Map showing plate boundaries and volcanoes.

operating at depth within subduction zones and which give rise to magmatism. These observations are usefully posed as a set of simple derivative questions.

1. Why does melting take place where cold lithospheric plates sink into and cool the mantle?

Subduction of cold oceanic lithosphere cools the surrounding mantle; thus it is somewhat paradoxical that subduction zones are sites of magmatism, volcanism, and high heat flow. It is now generally accepted that volatiles, in particular H_2O, play an important role in this regard by inducing partial melting and magma generation in subduction zones through the effects which H_2O has on lowering the solidus temperature of mantle peridotite and of the subducted oceanic crust. However, many unanswered questions arise therefrom, such as the origin of these volatiles and whether the principal source of magmas is the subducting oceanic crust or the overlying mantle wedge. Moreover, at what temperatures are subduction zone magmas produced? In addition to H_2O, is there a source of heat which contributes to the generation of magma and the origin of high surface heat flow observed in subduction zones?

2. What is the origin and significance of the spatial distribution of subduction zone volcanoes?

Subduction zone volcanoes tend to be concentrated in arcuate volcanic chains, which lie parallel to the deep ocean trenches that mark the surface plate boundaries and onset of subduction. Volcanism is restricted to the overriding lithospheric plate and usually to sites where the depth to the underlying subducting plate is ~100 km or more. The trenchward limit of volcanoes tends to form a well-defined line, called the *volcanic front*, which is consistently located at ~110 km above the subducting plate. Some subduction zones are characterized by double volcanic chains parallel to, but at different distances from, the trench. Subduction zone volcanoes also often exhibit regular spacing along a volcanic arc, as, for example, the constant spacing occurring between the Cascade volcanoes: Mount Rainier, Mount St. Helens, Mount Adams, Mount Hood, Mount Jefferson, Three Sisters, and so on. These spatial patterns of volcano distribution are clearly a reflection of the underlying regions of magma generation within subduction zones.

3. Why is the chemical composition of subduction zone magmas so distinct from magmas in other tectonic settings?

Numerous studies have demonstrated the distinct chemical characteristics of subduction zone magmas compared with other tectonic settings, differences which are often employed to discriminate tectonic settings for old magmatic/volcanic rocks. What are the processes unique to the generation of subduction zone magmas that result in these distinct geochemical characteristics?

4. Why does the chemical composition of magmas often display systematic change across volcanic arcs?

A feature of many, though not all, subduction zones is the occurrence of systematic variation in magma chemistry across a volcanic arc. This is also known as the *K-h* or Dickinson-Hatherton relationship, whereby the K content of magmas increases with height above the subducting lithospheric plate and distance from the trench axis. We shall draw together evidence supporting a general/global trend in across-arc chemical variation of this nature. We also evaluate the processes producing these systematics and investigate the nature of exceptions and their origins.

1.2 Petrogenetic Models for Subduction Zone Magmatism

Models and modeling are integral to the study of the origin of subduction zone magmatism, as the magma generation process occurs at depths beyond that of direct observation. The results obtained from forward and inverse modeling strategies applied to geophysical and geochemical data in experimental petrology, trace element and isotope geochemistry, seismology, and others form the foundation for general petrogenetic models of subduction zone magmatism.

1.2.1 Existing models

There currently exists a spectrum of petrogenetic models for subduction zone magmatism. These models can be divided into two broad groups, depending on whether the subducting oceanic crust or the mantle wedge is perceived to be the principal source of magmatism.

Models which have invoked melting of the subducting oceanic crust have been proposed, among others, by Holms (1965), Ringwood and Green (1966), Nicholls and Ringwood (1973), Wyllie and Sekine

(1982), and Brophy and Marsh (1986). The model proposed by Wyllie and Sekine (1982), based on phase equilibria in the model system $KAlSiO_4$-Mg_2SiO_4-SiO_2-H_2O, is a particularly informative example of this type (Figure 1.3), and bears similarities to the hybridization process first proposed by Nicholls and Ringwood (1973). In Wyllie and Sekine's model, hydrous silica-rich magmas formed by partial melting of the basaltic oceanic crust at depth rise into the overlying mantle wedge where they react to produce zones of phlogopite, orthopyroxene, and clinopyroxene within the base of the mantle wedge. The H_2O originally dissolved in the hydrous magma cannot be incorporated completely into the solid hydrous phases formed (i.e., phlogopite), so it continues to ascend as a separate fluid phase, able to cause partial melting of peridotite at higher and hotter regions within the mantle wedge. Wyllie and Sekine (1982) and, later, Wyllie (1988) emphasized the vertical relationship between these separate magma generation regions in the subducting oceanic crust and mantle wedge (M and N in Figure 1.3) and their location at and above the point where the subducting basaltic ocean crust exceeds its solidus temperature. They further proposed that the hybridized phlogopite-pyroxene

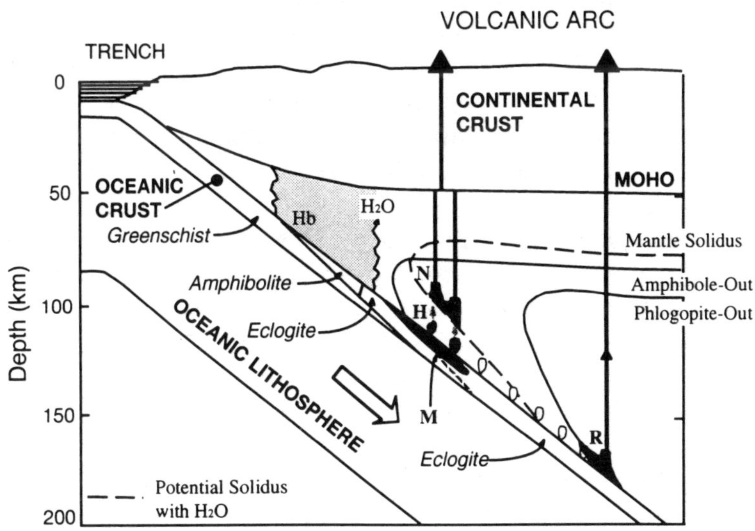

Figure 1.3 A petrogenetic model for subduction zones. Locations of partial melting (M,N,R) and metasomatism (H) are shown. *(Reproduced by permission from Wyllie and Sekine, 1982.)*

bodies at the base of the mantle wedge are dragged downward above the subducting plate until the solidus temperature of phlogopite is exceeded, at which point further melting takes place to produce magmas of alkaline (K-rich) composition. This model explains the development of the volcanic front, the occurrence of dual volcanic chains in some subduction zones, and the development of across-arc variation in magma chemistry.

Models which invoke the mantle wedge as the principal source of subduction zone magmas (e.g., Coats, 1962; McBirney, 1969; Ringwood, 1975; Kushiro, 1983, 1987; Tatsumi et al., 1983, 1986; Nye and Reid, 1986; Sakuyama and Nesbitt, 1986; Gust and Perfit, 1987) have emphasized the role of subsolidus dehydration of the subducting oceanic lithosphere and the release of H_2O into the overlying mantle wedge, where melting is induced by the drastic lowering of the solidus temperature of peridotite in the presence of H_2O (down from at least 1300°C to ~1000°C at 1.0 GPa; Kushiro, 1969). Most models in this category have assumed the pervasive supply of H_2O to the mantle wedge, such that melting would occur everywhere in the mantle wedge where the temperature exceeds ~1000°C. In contrast, Sakuyama and Nesbitt (1986) considered that dehydration of the subducted oceanic lithosphere was likely to be confined largely to beneath the forearc and that water was not supplied directly from the subducting plate to the magma source in the mantle wedge. Instead, they proposed that hydrous amphibole-bearing peridotite is formed by reaction with subducting plate-derived H_2O beneath the forearc and is dragged downward above the subducting plate until it undergoes partial melting at the solidus temperature (~1100°C) of H_2O-undersaturated, amphibole-bearing peridotite. An integral part of this and similar models of the same category is that the distinct geochemical characteristics of subduction zone magmas are created by the transfer of elements which are soluble in a H_2O-rich fluid phase from the subducting oceanic lithosphere to the mantle wedge magma source. These models also provide the basis for suggesting that the development of a partially molten layer near the base of the mantle wedge and beneath denser unmolten peridotite presents a potential source of partially molten instabilities that may ascend in the form of diapirs through the mantle wedge. Working within this framework, Sakuyama and Nesbitt (1986) postulated that across-arc variation in incompatible element abundances could be attributed to decreasing degrees of partial melting away from the volcanic front due to deeper melt segregation from mantle diapirs.

1.2.2 Toward a consistent model

Many existing models usually account for most but not all of the salient features of subduction zone magmatism. The relatively sophisticated models of Sekine and Wyllie (1982) and Sakuyama and Nesbitt (1986) are no exception, for they relate the onset and positioning of melting to a particular temperature. However, temperature distributions are likely to be highly variable from subduction zone to subduction zone (e.g., Peacock, 1990), and thus confront these models with a need to account for the observed constancy in depth (i.e., ~110 km) to the subducting plate beneath the volcanic front in most subduction zones.

At the same time it is also fair to say that there exists such a large range of physical and chemical variables in subduction zones that no single model is likely to be able to account for all observations in all subduction zones. Likewise any model developed on the basis of observations in any single subduction zone may not apply universally. On the other hand, if a case can be made for certain observations being generally applicable to all or many subduction zones, then these should be incorporated into the foundation for a consistent general model. This has been our approach, and the general observations posed as questions in Section 1.1 are a selection which we perceive to apply at a first-order level. The construction of a consistent model about these and additional detailed geophysical and geochemical observations is the aim of this book. The basic framework about which we develop such a model is illustrated schematically in Figure 1.4 and is similar to that proposed by Tatsumi (1989). The principal elements and outcomes of this particular model are as follows.

1. Melting of the mantle wedge is induced by influx of H_2O released from pressure-sensitive dehydration reactions in hydrous peridotite as it is dragged downward by subduction along the base of the mantle wedge.

The subducting oceanic lithosphere may be effectively anhydrous beneath the volcanic arc due to its dehydration having approached completion beneath the forearc; hence, the subducting lithosphere can be considered unable to supply H_2O directly to the mantle wedge

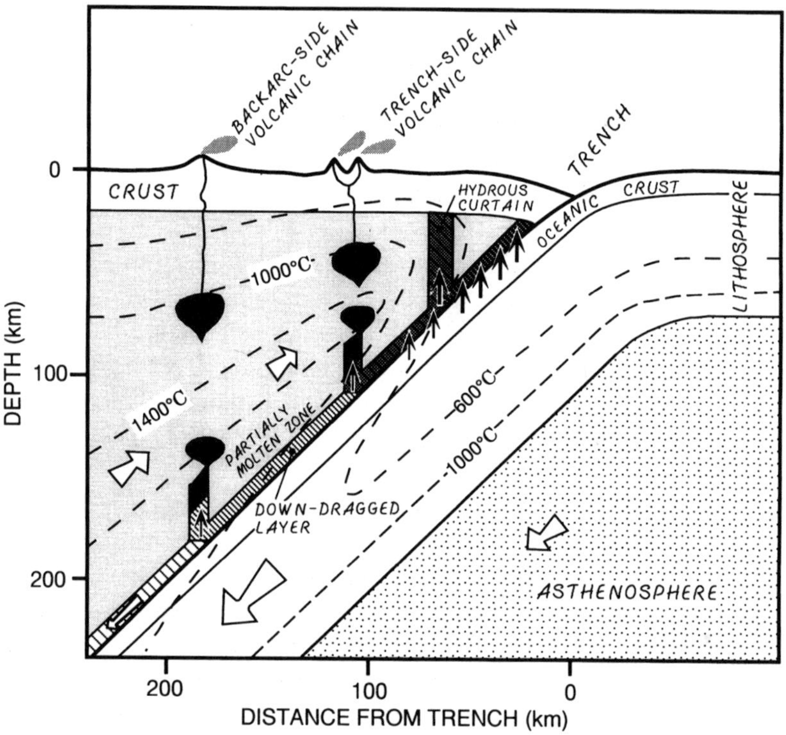

Figure 1.4 A possible model for subduction zone magmatism.

source beneath the volcanic front. However, hydrous peridotite formed within the forearc mantle wedge, through interaction with subducting lithosphere-derived H_2O, can release H_2O to the mantle wedge beneath the volcanic front as it is dragged downward by viscous coupling to the subducting plate. Dehydration reactions in this hydrous peridotite layer at the base of the mantle wedge are likely to be concentrated at three different depths, which correspond to the decomposition of (1) serpentine at a temperature of ~600°C, (2) amphibole/chlorite at a pressure of 3.5 GPa (~110-km depth), and phlogopite at ~6.0 GPa (~200-km depth). H_2O released by these dehydration reactions will ascend into the mantle wedge in the form of rising "hydrous curtains." The solidus temperature of hydrous peridotite may not be exceeded in the forearc mantle wedge, except in circumstances where unusual higher temperature conditions exist, in which case near-trench magmatism of high-Mg andesite or boninite composition may take place. Under normal conditions, the solidus temperature of hydrous peridotite is attained beneath the volcanic arc and at deeper levels within the mantle wedge, so partially molten zones form in the mantle wedge above the amphibole/chlorite and phlogopite dehydration reactions at 3.5 and ~6.0 GPa.

2. Preferential transport of elements soluble in H_2O-rich fluids gives rise to the distinctive geochemical characteristics of subduction zone magmas and their mantle wedge source regions.

Experimental data and theoretical considerations provide evidence for enhanced solubility of trace elements with larger ionic radii and smaller charge (i.e., higher ionic potential) in aqueous fluids at upper mantle conditions. The characteristic enrichments of the large-ion lithophile or K-group elements (Cs,Rb,K,Ba,Sr) and depletions of high-field-strength elements (Nb,Ta,Zr,Hf,Ti) that have been documented for subduction zone magmas are consistent with the preferential transport of more fluid-soluble elements into the mantle wedge source region. Aqueous fluids released from dehydration reactions occurring in both the subducted oceanic lithosphere and the downdragged hydrous layer therefore play a key role in inducing melting and in producing the diagnostic chemical characteristics of subduction zone magmas.

3. Secondary convection induced by subduction produces high-temperature conditions in the mantle wedge.

Addition of H_2O to the mantle wedge is certainly essential to the instigation of partial melting; however, this may not in itself be sufficient to explain observed high heat flow in volcanic arcs and the occurrence of arc basalt magmas with liquidus temperatures of 1300°C or more. An additional source of heat in the mantle wedge is required. This may be accounted for by convective flow in the mantle wedge induced by viscous coupling to the subducting plate, which causes high-temperature mantle materials to be drawn into the shallow mantle wedge from the asthenospheric upper mantle.

4. Variation in the depth of magma segregation from mantle diapirs, possibly corresponding to variable depth to the lithosphere/asthenosphere boundary, may result in differing degrees of partial melting and determine overall incompatible element concentrations in subduction zone magmas.

Mantle diapirs initiated from the partially molten zone at the base of the mantle wedge may rise through the mantle wedge until their ascent is arrested near or at the lithosphere/asthenosphere boundary, where the degree of melting will reach a maximum, and incompatible trace element concentrations a minimum. The rheological boundary or transition between the asthenosphere and lithosphere may be associated with a particular mantle wedge isotherm and located at a deeper level beneath the backarc side of a volcanic arc. This possibility is consistent with modeling of induced convection within the mantle wedge and the general observation that incompatible trace element concentrations increase in magmas away from the volcanic front.

2 Physical Characteristics of Subduction Zone Magmatism

2.1 Geophysical Observations

Subduction zones are tectonically unique locations where the oceanic lithosphere sinks into the mantle. At the surface these zones are usefully subdivided into three regions: the forearc, volcanic arc, and backarc (Figure 1.1). The forearc is bounded by the trench, a deep depression in the surface topography that marks the surface plate boundaries and onset of subduction, and the volcanic front, the line formed by those volcanoes nearest and usually parallel to the trench (Sugimura, 1960; Figure 2.1). In some subduction zones the forearc portion nearest the trench consists of a steeply inclined, fault-bounded wedge of sediments and volcanics, scraped off the top of the subducting ocean crust, and commonly called either an accretionary prism or subduction zone complex. The volcanic arc is the region where subduction-related volcanism takes place. The backarc extends from beyond the volcanic arc, and in some subduction zone systems is the site of rifting and magmatism, in a style with similar physical and chemical features to that at mid-ocean ridges (Figure 1.1).

The internal structure of subduction zones is accessible via a diversity of geophysical methods and observations. The location and structure of the subducted lithospheric plate, the mantle wedge and volcanic arc crust, as well as the thermal structure and the presence of melt are some of the features able to be discerned. These and other geophysical characteristics of subduction zones shall be outlined herein using an EW cross section through the NE Japan arc at 40°N (Figure 2.2), one of the few cross sections for which high-resolution

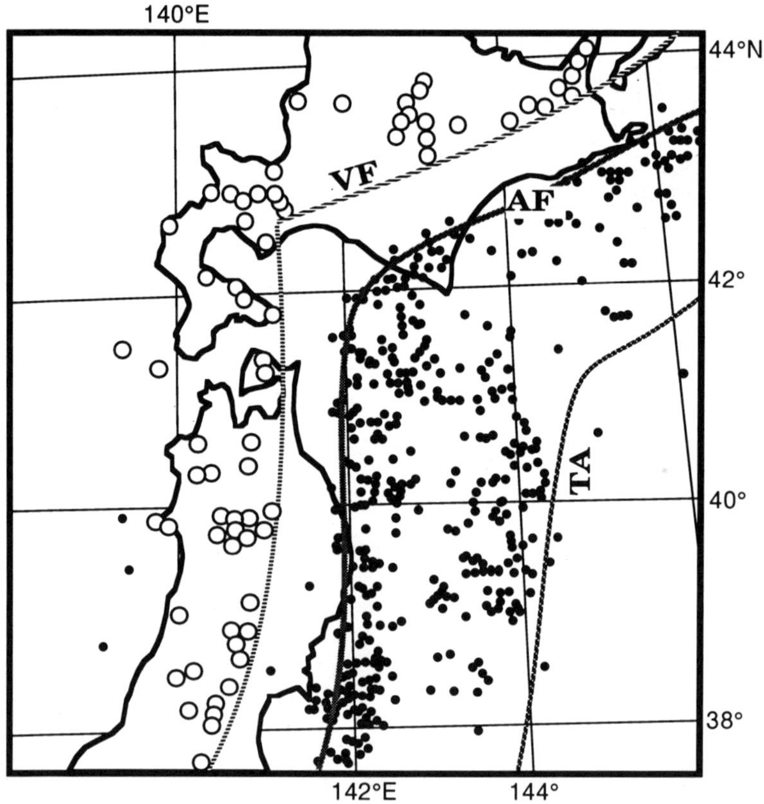

Figure 2.1 Distribution of Quaternary volcanoes (open circles) in the NE Japan arc and earthquakes occurring at depths between 40 and 60 km in the mantle (filled circles). The volcanic front (VF) marks the line of volcanoes nearest the trench axis (TA), and the aseismic front (AF) is the boundary between the seismogenic mantle and aseismic mantle in the forearc region.

geophysical data sets are available. Important variations upon the NE Japan arc case or instances where it is atypical shall be illustrated as appropriate with examples from other subduction zones.

2.1.1 Topography

The NE Japan forearc slopes smoothly down toward the trench. No forearc basin or depositional depression is developed as is sometimes the case, particularly in continental arcs, such as the Sunda arc where a marked topographic depression occurs along the strike length of the forearc for a distance of ~700 km (Moore et al., 1982).

The NE Japan volcanic arc forms an emergent topographic high, which may be compared to completely submerged cases where very young volcanic arcs are built on oceanic crust through to completely emergent cases where mature volcanic arcs are developed on thickened crust, typically at continental margins.

The NE Japan backarc slopes more steeply than the forearc and changes abruptly from volcanic arc crust to the thinner oceanic crust which floors the Japan Sea, a backarc basin. Backarc basins are commonly developed behind volcanic arc-trench systems, most notably along the W. Pacific margin, with exceptions being limited largely to continental margin volcanic arcs such as Alaska and Chile. The absence or presence of backarc basins behind volcanic arc-trench systems has been used to divide subduction zones into Chilean (absent) and Mariana (present) types of Uyeda and colleagues (Uyeda and Kanamori, 1979; Uyeda, 1979, 1982, 1986). Chilean-type subduction zones also tend to have relatively high stress, shallow-dipping subduction of relatively young lithosphere, relatively large and shallow earthquakes, a large accretionary prism, and intermediate to felsic volcanism, whereas Mariana-type subduction zones have relatively low stress, steep-dipping subduction of relatively old lithosphere, few large earthquakes, little or no accretionary prism, and mafic to intermediate volcanism. This two-end-member classification scheme is appealing in its simplicity; however, analysis of sediment accretion (von Huene and Scholl, 1991) and sesmic coupling (Pacheco et al., 1993) reveals that they are complex variables which do not relate simply to the presence or absence of a backarc basin. The origin of backarc basins will be discussed in Chapter 7.

2.1.2 Gravity

Gravity data provide constraints on the internal density structure of subduction zones, with Bouguer and free-air gravity anomalies being employed for land and sea areas, respectively.

Most volcanic arc-trench systems display (1) a negative anomaly at the trench, reaching a minimum immediately landward of the trench, (2) a positive anomaly associated with the volcanic arc, and (3) a positive anomaly in the region immediately oceanward of the trench (see Figure 2.2). These characteristics are consistent with the subduction of relatively dense oceanic lithosphere beneath the volcanic arc, beginning from the position of the trench, and the development of a

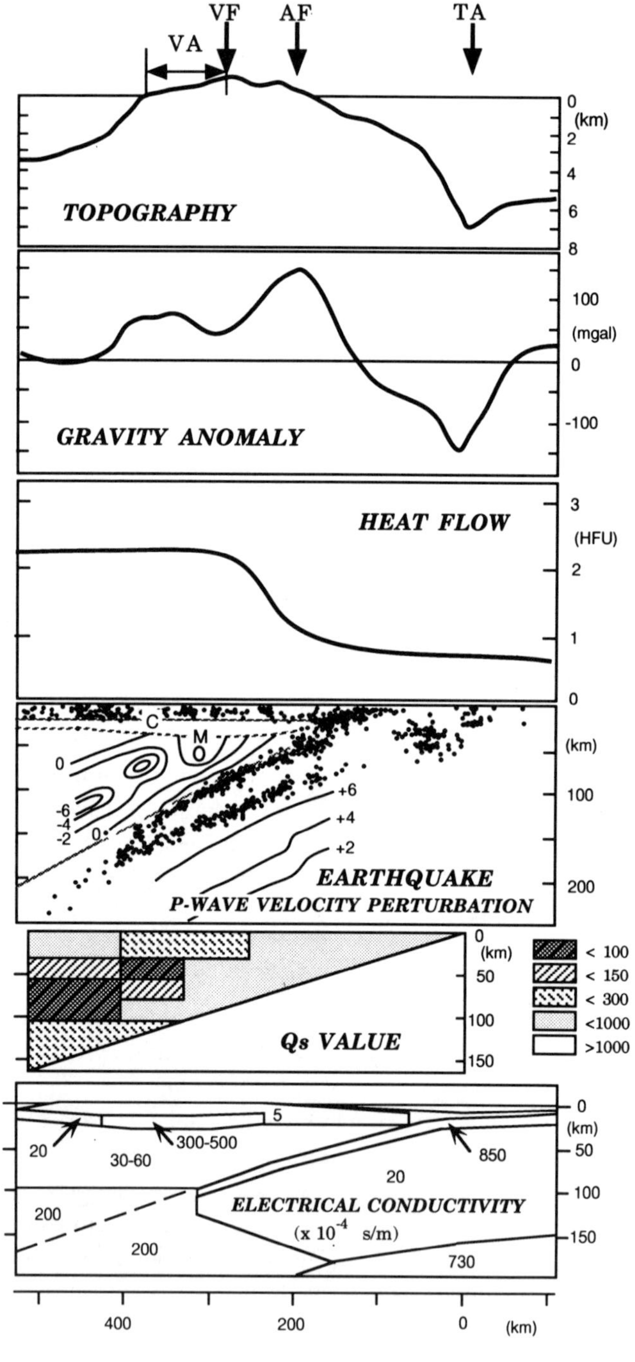
VF
AF
TA
VA
TOPOGRAPHY
0
(km)
2
4
6
8
GRAVITY ANOMALY
100
(mgal)
0
-100
HEAT FLOW
3
(HFU)
2
1
0
C
M
O
0
-6
-4
-2
0
+6
+4
+2
(km)
100
200
EARTHQUAKE
P-WAVE VELOCITY PERTURBATION
Qs VALUE
0
(km)
50
100
150
< 100
< 150
< 300
<1000
>1000
ELECTRICAL CONDUCTIVITY
(x 10^-4 s/m)
5
20
30-60
300-500
850
20
200
200
730
0
(km)
50
100
150
400
200
0
(km)

topographic high due to upward hinging of the subducting lithospheric plate outboard from the trench.

2.1.3 Heat flow

Surface heat flow data enable constraints to be placed on the thermal state and structure of the Earth's interior, given knowledge of the thermal gradient and thermal conductivity of the underlying materials.

In the NE Japan arc and other volcanic arc-trench systems, a number of characteristic heat flow features have been demonstrated by Uyeda and Horai (1964), Watanabe et al. (1977), Blackwell et al. (1982), Ziagos et al. (1985), Stern (1987), and Henry and Pollack (1980), among others, including (1) relatively low heat flow (< ~1 HFU, heat flow unit) in the forearc region, and (2) variable but higher heat flow (>2–3 HFU) in the volcanic and backarc regions (Figure 2.2). These "normal" heat flow observations can be explained in one of two ways: (1) shear heating along the interface between the subducting oceanic plate and the base of the mantle wedge (e.g., McKenzie and Sclater, 1969) or (2) heat influx due to subduction-induced convection in the mantle wedge (e.g., Minear and Toksöz, 1970).

Shear heating along the subducting plate–mantle wedge interface may occur due to resistance to the subduction of the lithospheric plate. The heat flux, q, produced by shear heating can be expressed approximately by

$$q \cong u\tau$$

where u and τ are the rate of subduction and shear stress, respectively. Early numerical simulations of the thermal structure in subduction

←

Figure 2.2 Geophysical cross sections and profiles for the NE Japan arc at 40°N. The top three panels show topography, gravity, and surface heat flow profiles across the arc. The fourth panel shows P-wave velocity perturbation in percent values (negative is slow and positive is fast) and the cross-sectionial distribution of earthquakes. Note the slow region inclined upward in the mantle wedge to beneath the volcanic front, the double-dipping seismic plane in the subducting lithospheric plate, and limitation of seismic events to above the Conrad discontinuity (dashed line marked C) within the volcanic arc crust, above the Moho (dashed line marked M). The fifth panel illustrates the Qs factor, a parameter indicating seismic attenuation, distribution in the overriding plate and mantle wedge. The bottom panel shows the electrical conductivity cross section.

zones (e.g., Hasebe et al., 1970; Minear and Toksöz, 1970; Toksöz et al., 1971) suggested that several kilobars (hundreds of megapascals) of shear stress were necessary to produce the normal pattern of heat flow observed in subduction zones. However, shear heating is an intrinsically self-limiting process because the heat produced lowers the viscosity and thus the shear stress experienced (Yuen et al., 1978), a feature not taken into account in early thermal structure modeling.

The alternative mechanism for explaining the general heat flow pattern is through heat advected into the mantle wedge, via convection induced therein by viscous coupling between the subducting oceanic lithosphere and the mantle wedge. This mechanism has been the subject of increasingly detailed investigations, which have taken into account the additional effects of endothermic dehydration reactions in the subducted oceanic crust and viscous shear heating in the mantle wedge (e.g., Andrews and Sleep, 1974; Bodri and Bodri, 1978; Toksöz and Hsui, 1978; Honda, 1985; Furukawa, 1993). These simulations produce relatively high temperature regions within the mantle wedge; however, their ability to match the high-temperature conditions required for the generation of subduction zone basaltic magmas remains in question, an aspect examined in Chapter 7.

In many continental margin subduction zones, the thickness of the volcanic arc crust is more than 30 km and requires that the effects of crustal heat generation also be taken into account when assessing heat flow observations. Furukawa and Uyeda (1989) estimated a value of 30 mW/m^2, equivalent to ~0.7 HFU, for the heat flow component generated within the crust of the NE Japan arc.

An interesting exception to the normal heat flow characteristics observed in subduction zones is the Nankai Trough (Yamano et al., 1984), where high heat flow (~3 HFU) is observed in the forearc. This anomalous situation can be related to relatively recent (17 Ma; Kobayashi and Nakada, 1978; Shih, 1980) rifting/opening of the Shikoku Basin and subduction of "hot" lithosphere beneath the Nankai Trough, with attendant expulsion of hot water from the sediments of the Shikoku Basin during the initial stages of oceanic plate underthrusting.

2.1.4 Seismic observations

Crustal Structure

The depths to the Moho and Conrad discontinuities beneath the NE Japan arc have been estimated by explosion seismology (e.g., Yoshii and Asano, 1972) and seismic tomography using three-dimensional inversion methods (e.g., Horiuchi et al., 1982a,b; Zhao et al., 1990). These studies demonstrate that the Moho lies 33 km beneath the volcanic arc and decreases in depth toward the backarc region. The occurrence of earthquakes is limited to the upper part of the crust (Brace and Byerlee, 1970; Eaton et al., 1970; Takagi et al., 1977), and the maximum depth of crustal earthquakes (~15 km) corresponds roughly to the depth of the Conrad discontinuity (Figure 2.2). Three mechanisms have been proposed to explain the absence of earthquakes in the lower part of the crust, the occurrence of (1) a brittle-ductile transition (Meissner and Strehlau, 1982; Sibson, 1982, 1984), (2) an unstable-sliding to stable-sliding transition (Tse and Rice, 1986), and (3) a change from unstable to stable plastic flow (Hobbs et al., 1986). Regardless of the mecha-nism the transition can be attributed to a temperature increase with depth in the crust, since the maximum depth of crustal earthquakes is observed to be inversely proportional to the terrestrial heat flow.

In the NE Japan arc, more than 98% of shallow earthquakes occur at depths shallower than 15 km. However, a small population of microearthquakes occur at depths of 25–40 km and have the following distinct characteristics (Takagi et al., 1977): (1) lower predominant frequencies (1.5–3.5 Hz) than earthquakes in the upper crust and the dipping seismic zones, (2) a maximum magnitude of 2.5, and (3) a restricted occurrence to beneath volcanoes. These low-frequency microearthquakes most likely reflect the migration of magmas through the lower crust and uppermost mantle.

Distinct S-wave reflectors within the shallow crust (10–15 km) have been documented in some volcanic areas (NE Japan: Horiuchi et al., 1988; Mizoue et al., 1982; Mammoth Lake, California: Peppin, 1987; Peppin et al., 1989). These reflectors are believed to correspond to the upper boundaries of magma reservoirs, due to their high reflectivity and restricted occurrence to beneath volcanoes (Sanford et al., 1973; Hasegawa et al., 1991).

A schematic cross section of the crust and uppermost mantle structure beneath the NE Japan arc is presented in Figure 2.3, based on the seismic observations of Hasegawa et al. (1991)

Mantle Wedge Structure

Heterogeneous seismic structure in the upper mantle of subduction zones was first demonstrated by Utsu (1967) for the NE Japan arc

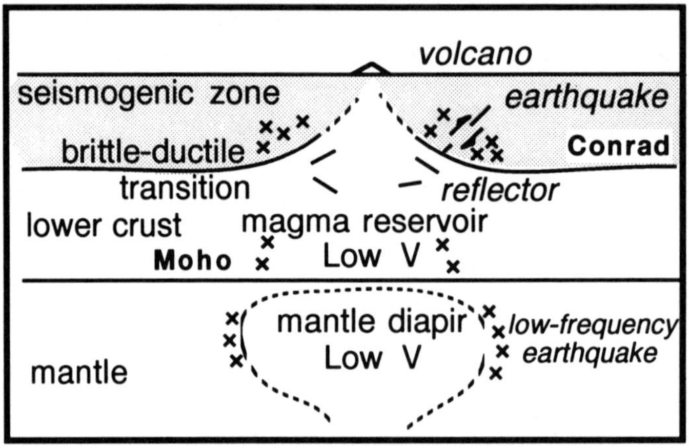

Figure 2.3 A schematic cross section of the crust and uppermost mantle beneath a volcano in the NE Japan arc. *(Reproduced by permission from Hasegawa, Zhao, Hori, Yamamoto and Horinouchi, 1991.)*

and by Oliver and Isacks (1967) for the Tonga arc. This heterogeneity includes an inclined high-velocity (V) and low-attenuation, high-*Q* zone (where *Q* is an efficiency factor for seismic-wave propagation that is inversely related to the attenuation of wave energy), sandwiched by low-V, low-*Q* regions on both the volcanic arc and oceanic sides. The inclined high-V, high-*Q* zone corresponds to the Wadati-Benioff zone and the position of the subducting oceanic lithosphere. The *Q* structure of the mantle wedge has been studied in several volcanic arc-trench systems, including Tonga (Barazangi and Isacks, 1972), the Andes and Kuriles (Sacks and Okada, 1974), and NE Japan (Hasegawa et al., 1979). The detailed structure of the latter is shown in Figure 2.2. These observations indicate the presence of a relatively hot region within the mantle wedge, the temperature of which has been estimated by Sato (1992) to reach a maximum of 1000–1100°C, based on the experimentally determined seismic properties of upper mantle peridotites (Murase and Kushiro, 1979; Murase and Fukuyama, 1980; Kampfmann and Berckhemer, 1985; Sato et al., 1988, 1989) and the seismic data of Umino and Hasegawa (1984). However, this maximum temperature estimate is insufficient to match the mantle wedge temperatures in excess of 1300°C which are required to produce basaltic magmas in the NE Japan arc (Tatsumi et al., 1983).

Seismic tomography studies using the inversion technique first described by Aki and Lee (1976) have in recent years provided detailed views of three-dimensional structures within the mantle wedge. Hasemi et al. (1984) have defined a fine-scale three-dimensional velocity structure in the mantle wedge of the NE Japan arc and have documented low-V columns immediately beneath several active volcanoes. A new tomographic method that enables high-resolution imaging of the P-wave velocity structure of the mantle wedge has been developed by Zhao (1991) (Figure 2.2), and has revealed that the low-V zone located immediately beneath Quaternary volcanoes in the NE Japan arc dips to the west in the mantle wedge and extends to a depth of ~150 km. A similar low-V structure has also been documented in other subduction zone systems (e.g., central Japan: Hirahara et al., 1989; S. Peru: Cunningham et al., 1986; New Zealand: Kuge and Satake, 1987; Tonga-Kermadec: Zhou, 1990; Aleutian: Engdahl and Gubbins, 1987).

The phenomenon of shear-wave splitting, whereby a shear wave passing through an anisotropic elastic medium splits into two orthogonally polarized components with different propagation speeds, can be used to measure the anisotropy along the wave path (e.g., Crampin, 1978). Ando et al. (1983) first observed shear-wave splitting within the mantle wedge beneath central Japan and proposed an origin due to melt-filled anisotropic cracks within the mantle. Similar shear-wave splitting effects in the mantle wedge have also been documented beneath Tonga (Bowman and Ando, 1987) and the northern Andes (Savage et al., 1989).

A final point of interest is that seismogenic mantle exists only beneath the forearc region as shown in Figures 2.1 and 2.2. The boundary between the seismogenic and aseismic mantle regions is called the aseismic front (Yoshii, 1975), and is located trenchward of the volcanic front.

Subducting Plate Structure

The dipping seismic zone (or Wadati-Benioff zone) is a characteristic feature of subduction zones (Benioff, 1954; Wadati, 1955). It has become well accepted that this seismic zone occurs within the high-V, high-*Q* region of the subducted oceanic lithosphere (e.g., Isacks et al., 1968). Accordingly, the structure of the downgoing slab has been estimated mostly from the distribution of deep earthquake foci (e.g.,

Isacks and Molnar, 1971); however, improved seismic networks and accumulated microearthquake data have now enabled improved resolution of the structure of the subducting plate. One of the most significant observations is the occurrence of two distinct, almost parallel, dipping seismic planes. A double-planed deep seismic zone was first confirmed by Umino and Hasegawa (1975) for the NE Japan arc and has since been documented in a number of other subduction zones (e.g., SW Kurile: Suzuki et al., 1983; Tonga: Kawakatsu, 1985; Alaska: Reyners and Coles, 1982).

The focal mechanisms of earthquakes along the double-planed seismic zone provide insights into the cause of these dipping seismic zones. Hasegawa et al. (1978) suggested, based on data from the NE Japan arc, that earthquakes with reverse faulting or with downdip compressional stress occur along the upper seismic plane, whereas downdip extensional stress occurs along the lower plane. No less than four processes have been proposed for this contrast in focal mechanism, including bending-unbending of the subducting plate (Engdahl and Scholz, 1977), thermal stress (Hamaguchi et al., 1983), sagging (Sleep, 1979), and phase transformations within the subducting plate (Veith, 1974). However, data sets from other subduction zones and more precise data from the NE Japan arc have revealed the occurrence of additional focal mechanisms to those proposed by Hasegawa et al. (1978), such that several different processes may need to be combined to explain the double-planed seismic zones, though bending-unbending of the subducted plate remains likely to play a principal role (Goto et al., 1985; Hasegawa and Takagi, 1987; Kawakatsu, 1986).

Phase conversion and wave refraction generally take place when seismic waves encounter discontinuities, and provide an additional source of information on deep structures within subduction zones. Okada (1971) first reported the forerunner of the ScS phase, S-wave reflection from the Earth's core at a station in Hokkaido, Japan, and demonstrated that the forerunner may be a P wave (ScSp) converting at the upper boundary of the subducting lithosphere. Numerous observations of this kind have been subsequently reported from various stations in Japan (Hasegawa et al., 1978; Okada, 1979; Nakanishi, 1980; Nakanishi et al., 1981) and the Andes (Snoke et al., 1977, 1979). P- to S- and S- to P-wave conversions across the top of the subducting plate have been used to estimate the location of the top of the plate

(NE Japan: Matsuzawa et al., 1986, 1987, 1990; Vanuatu: Chiu et al., 1985; Alaska: Reyners and Coles, 1982; Tonga: Mitronovas and Isacks, 1971). The upper boundary of the subducted slab beneath the NE Japan arc, as determined by Matsuzawa et al. (1986), is shown in Figure 2.2.

The efficient conversion between and the resemblance of ScS and ScSp phases led Okada (1979) and Nakanishi (1980) to suggest the presence of a low-velocity zone in the vicinity of the mantle wedge/ slab interface beneath Japan. This has been further documented in some arc-trench systems based on three-dimensional inversion methods (Hindu-Kush: Roecker, 1982; NE Japan: Miyamachi and Moriya, 1984; Taiwan: Roecker et al., 1987), PS-P time (NE Japan: Matsuzawa et al., 1986), and P and S later phases (SW Japan: Fukao et al., 1983; Hori et al., 1985; Hori, 1990). Matsuzawa et al. (1986) further demonstrated that the low-V layer has a thickness of less than 10 km near the top of the subducted plate at depths between 60 and 150 km, and that this layer includes the upper seismic plane (Figure 2.4). The P-wave velocity in the low-velocity layer was estimated to be ~7 km/s (Hori et al., 1985; Matsuzawa et al., 1986; Hurukawa and Imoto, 1987), which is comparable to that for the gabbroic layer of the oceanic crust (e.g., Christensen and Salisbury, 1975). This has led Fukao et al. (1983), Sacks (1983), and Hori et al. (1985) to the conclusion that the low-velocity layer is composed of gabbro/amphibolite, which is stable at

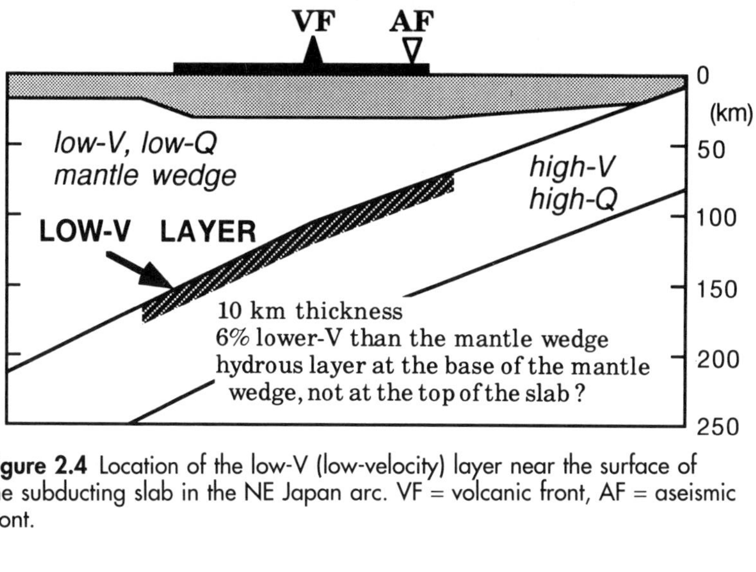

Figure 2.4 Location of the low-V (low-velocity) layer near the surface of the subducting slab in the NE Japan arc. VF = volcanic front, AF = aseismic front.

depths greater than the basalt/eclogite transition under low-temperature conditions. However, this explanation for the low-velocity layer, which exists to depths of at least 150 km beneath the NE Japan arc (Matsuzawa et al., 1986), is inconsistent with seismic coupling and focal mechanism observations that indicate the basalt/eclogite transformation probably takes place at depths shallower than 70 km (Ruff and Kanamori, 1983; Hurukawa and Imoto, 1992). An alternative interpretation of the low-velocity layer, is as a downdragged hydrous peridotite layer at the base of the mantle wedge.

Subducting Plate Penetration into the Deep Mantle

The question of whether the subducting oceanic lithosphere penetrates the lower mantle is of major significance to mantle convection, for penetration into the lower mantle requires whole-mantle convection rather than separate two-layered upper and lower mantle convection. In the 1970s, it was generally accepted that subducted plates did not penetrate the lower mantle (e.g., Richter, 1979), based on two lines of evidence: (1) there is no seismicity deeper than 700 km within the lower mantle (e.g., Richter, 1979), and (2) downdip compressional stresses are dominant for earthquakes in subducted plates at the upper/lower mantle boundary (e.g., Isacks and Molnar, 1971).

The possibility that subduction penetrated into the lower mantle was first proposed by Jordan (1977) and has been a topic of vigorous debate. Creager and Jordan (1986) provided evidence for plate penetration into the lower mantle beneath W. Pacific subduction zones based on travel time residual-sphere studies, and Jordan et al. (1989) further argued that the residual-sphere data favor subducting plate penetration into the lower mantle to depths of at least 1000 km. Studies of amplitude anomalies (e.g., Silver and Chan, 1986) have also supported lower mantle penetration.

Another means of examining possible subducting plate penetration into the lower mantle is via seismic tomography with resolution analyses (e.g., Kamiya et al., 1988; Miyatake and Hirahara, 1989; Zhou and Clayton, 1990; Spakman et al., 1989; Inoue et al., 1990). Studies of this kind have clearly delineated the inclined high-V zone of the subducted lithosphere and its correspondence to the Wadati-Benioff zone and have been used to suggest that plate penetration occurs at least in the Kurile arc. However, tomography for the Mariana arc by Zhou and Clayton (1990), together with three-dimensional ray-tracing analysis (Okino et al., 1989), indicate a horizontally inclined slab at the base of the upper mantle. This stands in conflict with the interpretation made by Creager and Jordan (1986), who have argued for lower mantle penetration in this region.

The possibility that lower mantle penetration may occur in some subduction zones but not others was suggested by Ringwood and Irifune (1988), who proposed that old cold slabs might penetrate the lower mantle, whereas young warm slabs might not, based on greater relative buoyancy of the latter (see Chapter 7). However, by itself, this reasoning does not account for the apparent absence of lower mantle penetration beneath the Mariana arc, where the oldest oceanic lithosphere on the Earth is being subducted.

2.1.5 Electrical conductivity

The electrical conductivity of mantle materials ranges from 10^{-7} s/m for subsolidus rocks to 10^{1} s/m for silicate melts, a contrast useful for probing the structure of the upper mantle and the presence of melt. Several underlying factors cause variation in electrical conductivity in the upper mantle.

(1) Electrical conduction is a thermal activation process; thus, the conductivity (σ) depends strongly on temperature (T):

$$\sigma = \Sigma\sigma_i \exp\left(\frac{-A_i}{KT}\right)$$

where σ_i is a constant for each constituent phase in the mantle, A_i is the activation energy required for generation and migration of charge carriers, and K is the Boltzmann constant. Shankland et al. (1981) and Omura et al. (1989) experimentally demonstrated the exponential increase of the electrical conductivity of olivine and basalt melts with increasing temperature.

(2) At equivalent temperatures, basaltic melts have two orders of magnitude greater electrical conductivity than mantle phases such as olivine (Shankland et al., 1981; Tyburczy and Waff, 1983).

(3) Electrical conductivity of a partially molten material depends on the connectivity of the partial melt (as does melt migration).

(4) The presence of conductive fluids such as H_2O may also enhance electrical conductivity because solid materials have comparatively low conductivities, especially at low temperatures where fluids may be present.

Conductivity measurements demonstrate that subduction of oceanic lithosphere with low electrical conductivity occurs in a more conductive mantle wedge (Figure 2.2). The high conductivity values in the mantle wedge are suggestive of relatively high temperatures in the region; however, the measured values (10^{-2}–10^{-3} s/m) remain below those expected for partially molten mantle (10^{0}–10^{-1} s/m) and indicate that large amounts of melting are not present in the mantle wedge (nor in the overlying volcanic arc crust).

The occurrence of a layer with higher electrical conductivity (~10^{-2} s/m) near the top of the subducted oceanic lithosphere has been documented in the NE Japan arc (Utada, 1988) and the Cordillera of North America (Waff et al., 1988; Wannamaker et al., 1989; Kurtz et al., 1990). The latter studies have suggested that the conductive zone at depths of 20 to 60 km lies above the surface of the subducted plate and have concluded that the conducting layer is probably due to fluid-filled cracks and pores associated with dehydration of the subducted oceanic crust.

2.1.6 Stress field

In most subduction zones, the tectonic stress field has a gradient in its horizontal component σ_H, with the vertical component σ_V remaining constant at the same depth and equal to the overburden pressure. The presence of a gradient in the horizontal stress field is typically manifested as an across-arc change in the nature and strike direction of fault systems, as shown schematically in Figure 2.5. Thrust faults typically occur in the forearc, strike-slip faults in the volcanic arc, and normal faults in the backarc (Nakamura and Uyeda, 1980).

Dyke swarms, faults, and earthquake focal mechanisms all reflect the orientation of the maximum horizontal stress, $\sigma_{H\max}$. Dykes generally propagate in a plane perpendicular to $\sigma_{\min}$, such that their strike represents $\sigma_{H\max}$. Accordingly, the surface expressions of dyke swarms, such as the distribution of monogenetic volcanoes upon or about a stratovolcano, may be used for determining the orientation of $\sigma_{H\max}$ (Nakamura, 1977; Zoback and Zoback, 1980). Although the displacement along faults is controlled essentially by the direction of strain, the maximum (or minimum) horizontal stress causing the faulting may approximate the $\sigma_{H\max}$ (or $\sigma_{H\min}$) orientation (Lensen and Otway, 1971; Pavoni, 1971). Earthquake focal mechanisms, estimated from the distribution pattern of push and pull for the primary displacement,

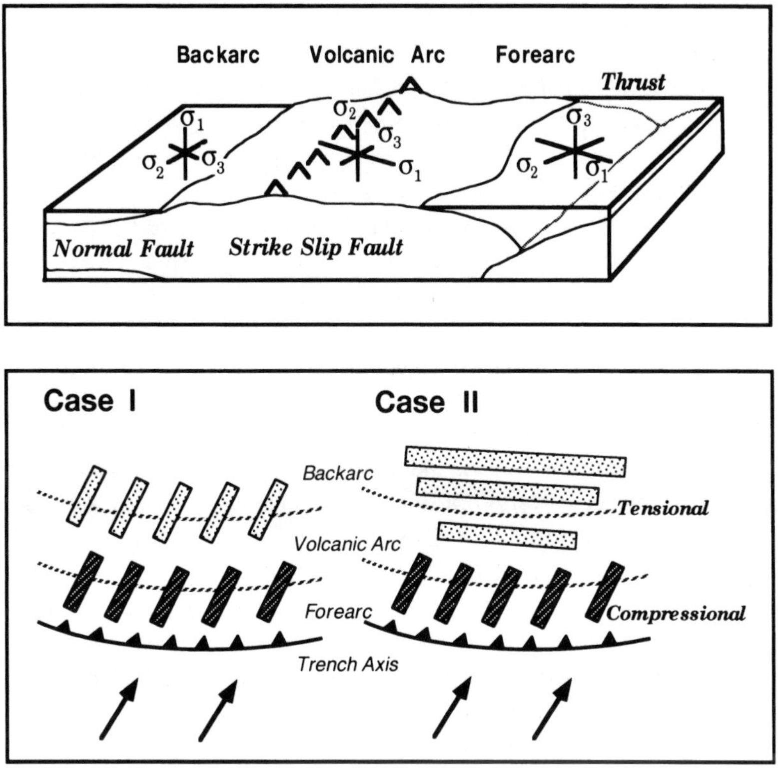

Figure 2.5 Typical subduction zone stress field variation (upper), with σ_H decreases toward the backarc and the two end-member paterns for σ_{Hmax} orientation in subduction zones (lower). *(Reproduced by permission from Nakamura and Uyeda, 1980.)*

can be used as an even more direct manifestation of stress state at their origin.

If we denote three principal stresses, σ_1, σ_2, and σ_3 ($\sigma_1 > \sigma_2 > \sigma_3$), then σ_{Hmax} can be σ_1 or σ_2. A compressional stress field is one where σ_{Hmax} is σ_1 and σ_{Hmin} is σ_2, and is characterized by the occurrence of thrust faulting. A tensional stress field occurs where $\sigma_{Hmax} = \sigma_2$ ($\sigma_{Hmin} = \sigma_3$, and $\sigma_V = \sigma_1$) and is characterized by normal faulting (Figure 2.5). Nakamura and Uyeda (1980) examined the trajectories of σ_{Hmax} in several arc-trench systems and proposed two end-member patterns of σ_{Hmax} trajectories (Figure 2.5). In one (Case I in Figure 2.5), the direction of σ_{Hmax} is almost parallel to that of plate convergence as σ_{Hmax} changes from σ_1 to σ_2, as observed in Central America, whereas in

Case II, $\sigma_{H\max}$ changes its direction from σ_1 to σ_2, as observed in the middle Miocene NE Japan arc, the Aleutian arc, and the Hellenic arc. Case I is explained by the gradual landward release of compressional stress caused by plate convergence, whereas Case II requires an additional force to act on the overriding plate to deflect the $\sigma_{H\max}$ axis, which may be a gravitational body force, shear stress due to induced convective currents in the mantle wedge, or due to forced injection of asthenospheric material into the mantle wedge (Nakamura and Uyeda, 1980; Tatsumi and Tsunakawa, 1992).

2.2 Spatial Distribution of Volcanoes

2.2.1 Two volcanic chains

Active volcanoes in subduction zones are not distributed evenly within volcanic arcs but tend to concentrate in certain areas or volcanic zones. The present distribution of volcanic zones in Japan is depicted schematically in Figure 2.6 and shows two volcanic chains forming subparallel zones in NE and central Japan. The occurrence of two volcanic chains within a single volcanic arc has been emphasized previously by Marsh (1979), examples of which include the Aleutian, Kamchatka, Kurile, NE Japan, Indonesia, and Scotia arcs. The evidence for these dual volcanic chains is summarized for a number of subduction zones in Figure 2.7, in the form of the frequency distribution of volcanoes as a function of distance from the volcanic front in each case. Details of these arc systems will be given.

Aleutian (Figure 2.8)

The Aleutian arc extends ~2500 km along the southern edge of the Bering Sea and Alaskan mainland, formed by subduction of the Pacific plate beneath the North American plate at the Aleutian trench. The orientation of subduction changes from normal to oblique from east to west along the arc. The volcanic arc comprises ~80 volcanoes and ends in the west at Buldir Island, where plate motion ceases to be convergent and becomes transform slip. The volcanic front is sharp and located close to the 100-km contour to the top of the subducted lithosphere (Kienle and Swanson, 1983; Jacob, 1986; Creager and Boyd, 1991). The volcanic front is segmented with discontinuities corresponding to the position of fractures/breaks in the subducted

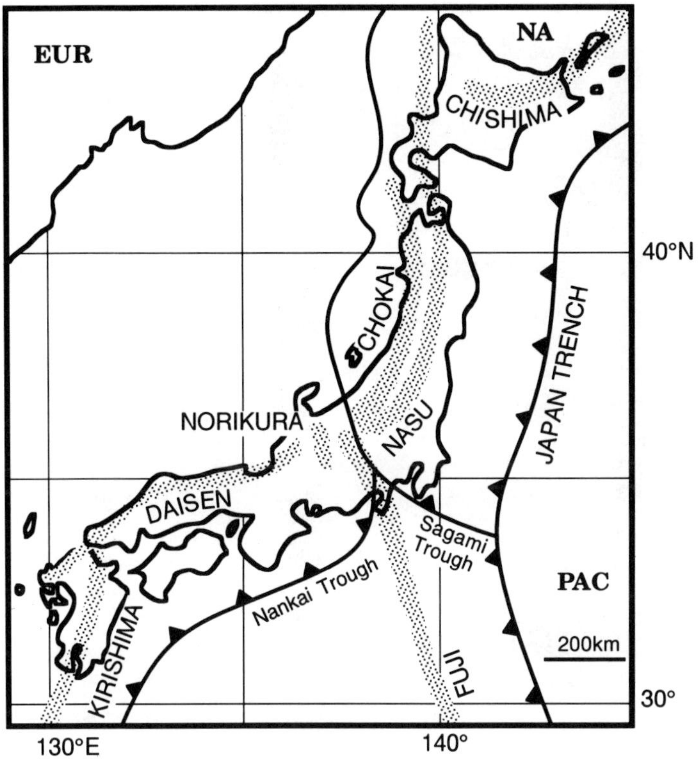

Figure 2.6 Distribution of volcanic zones in the Japanese Islands. Note the formation of two trench-parallel volcanic chains in the NE Japan arc, the Nasu and Chokai volcanic chains. PAC = Pacific plate, NA = North American plate, EUR = Eurasian plate.

Pacific plate (Marsh, 1979). Only two volcanoes, Bogoslof and Amok, occur behind the volcanic front, at a distance of 60–70 km in the eastern Aleutians (Figures 2.7 and 2.8). Basalts of late Tertiary and Quaternary age are scattered in the far backarc Beringia region of western Alaska (Figure 2.8); however, this particular volcanism may be associated with incipient continental rifting (Turner et al., 1981) rather than with subduction.

Kamchatka (Figure 2.9)

Kamchatka Peninsula is located at the northern end of the 2000-km-long Kurile arc, due to subduction of the Pacific plate at the Kurile Trench. Approximately 100 volcanoes form a trench-side volcanic

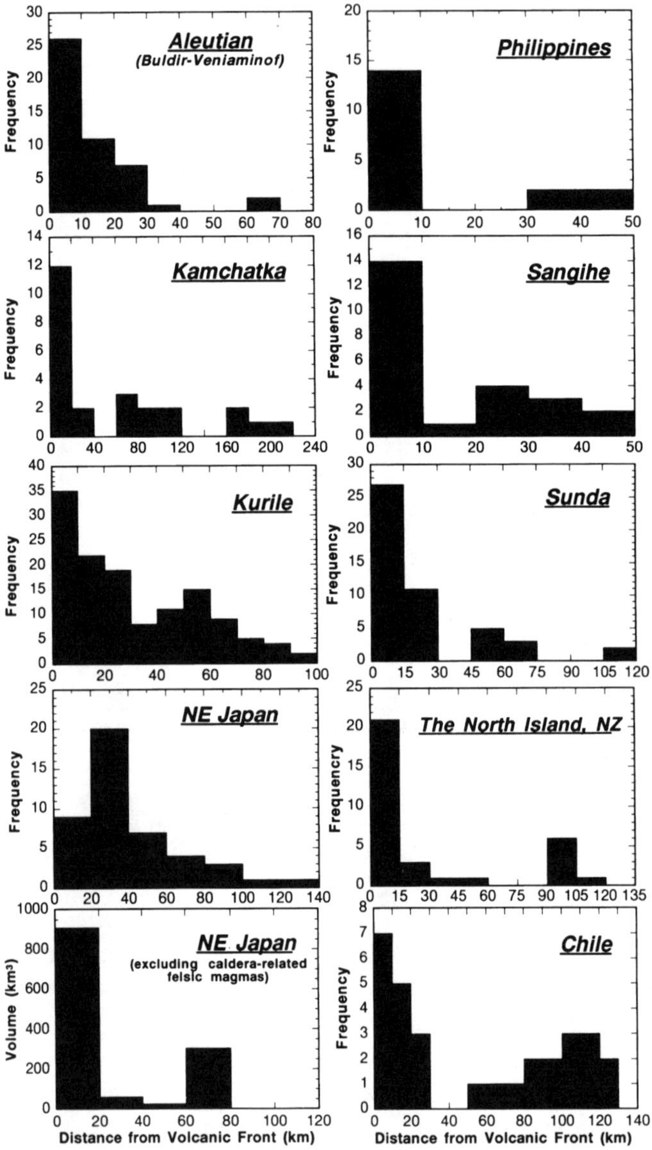

Figure 2.7 Frequency distributions for distances between individual volcanoes and the volcanic front in various volcanic arcs.

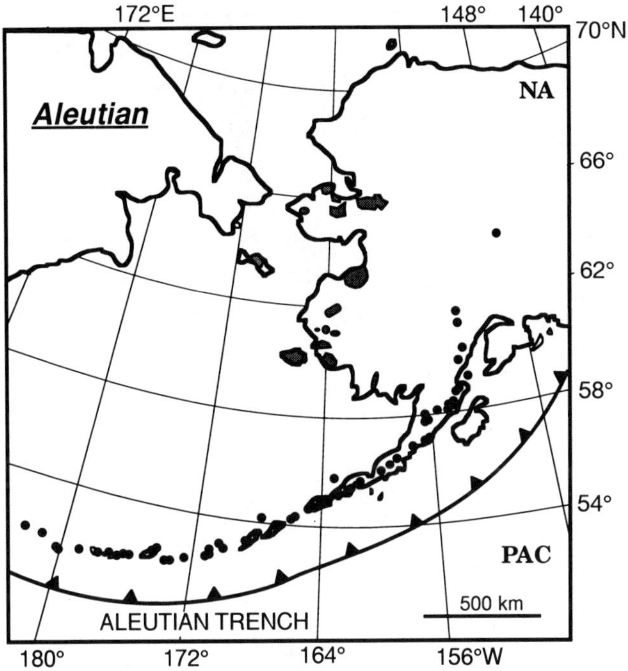

Figure 2.8 Distribution of volcanoes (solid circles) for the Aleutian arc. Hatched areas show localities of Neogene basalt fields. PAC = pacific plate, NA = North American plate.

chain, parallel to the trench axis and ~100–120 km above the surface of the subducting plate (Tarakanov and Kim, 1983). A second volcanic chain is formed between 60 and 120 km behind the volcanic front in the north and also in the south, and consists of about 20 volcanoes (Figure 2.9). Voluminous basalts, mostly forming lava plateaus, occur more than 200 km behind the volcanic front. These basalts are geochemically similar to those of the first and second chain lavas and may constitute an unusual example of a third volcanic chain (Tatsumi et al., 1994a), though no seismic zone is observed to extend beneath this third volcanic chain.

Central Kurile (Figure 2.10)

The Kurile Islands, located between Kamchatka and Hokkaido, comprise about 200 submarine and subaerial volcanoes. The volcanic front is segmented by a bend of 22–23° near Boussole Strait in the central

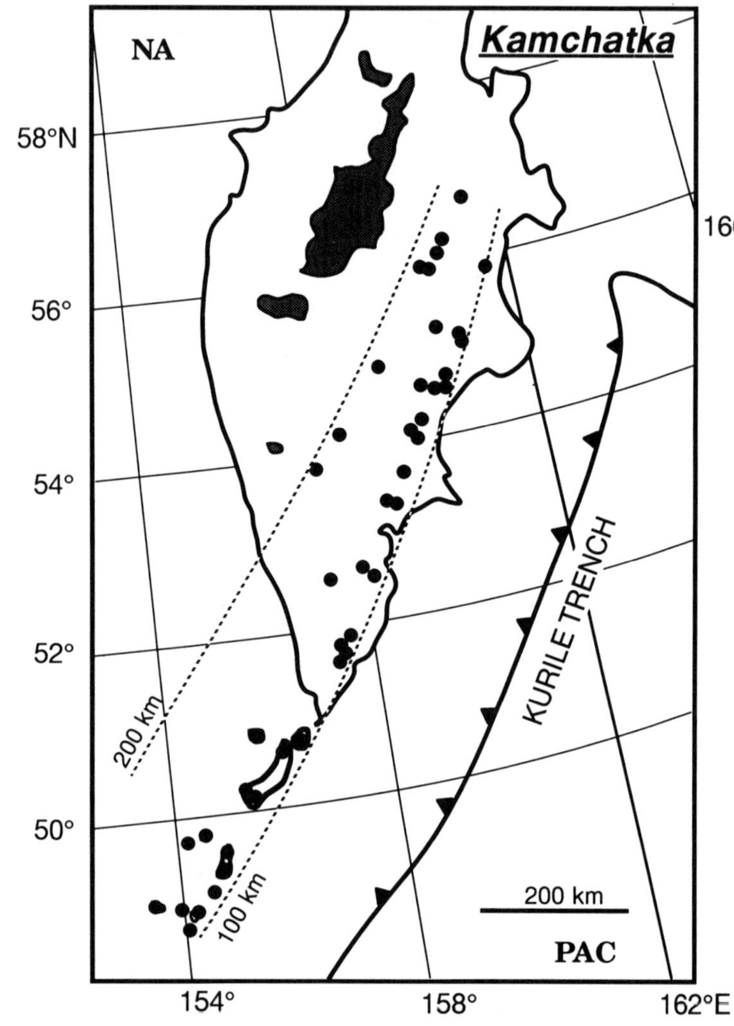

Figure 2.9 Distribution of volcanoes (solid circles) and depth contours (dashed lines) to the surface of the dipping seismic zone for Kamchatka. Stratovolcanoes (solid circles) form two volcanic chains at ~120 km and ~180 km above the dipping seismic zone, whereas plateau lavas (hatched fields) occur in the backarc region. PAC = pacific plate, NA = North American plate.

Kurile Islands (Avdeiko et al., 1991). A bimodal distribution of volcanoes across this volcanic arc has been demonstrated by Avdeiko et al. (1991), with the trench- and backarc-side volcanic chains located ~120 km and ~170 km above the dipping seismic zone, respectively

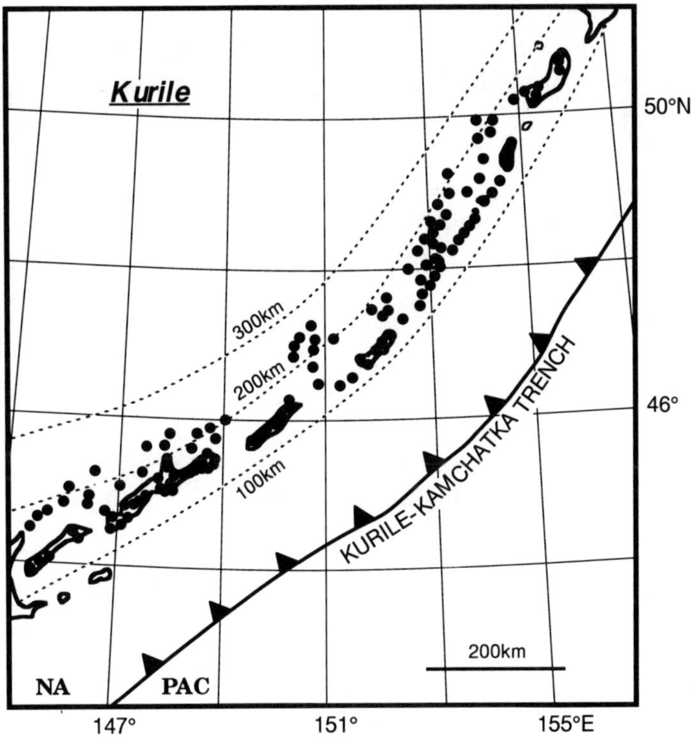

Figure 2.10 Distribution of volcanoes (solid circles) and depth contours (dashed lines) to the surface of the dipping sesimic zone for the Kurile arc. PAC = pacific plate, NA = North American plate.

(Tarakanov, 1987). More than 60% of the volcanoes are concentrated within the trench-side volcanic chain.

NE Japan (Figures 2.1 and 2.6)
The NE Japan arc forms a 700-km-long volcanic arc-trench system. Two parallel belts of volcanism occur, the Nasu zone in the east and the Chokai zone some 60 to 100 km farther west. The frequency distribution of volcanoes across this arc does not distinguish these volcanic chains, due to a clustering of volcanoes near 40°N within the Nasu trench-side chain (Figures 2.1, 2.6, and 2.7). However, the distribution of volume of volcanic material, excluding caldera-related felsic rocks, clearly reveals the occurrence of two volcanic chains (Figure 2.7), with the volcanic front located ~110 km above and the backarc-side chain ~150 km above the surface of the subducted litho-

sphere (Hasegawa et al., 1991). The number of volcanoes and volume of volcanic materials is greatest along the volcanic front and decreases exponentially toward the backarc side of the arc (Sugimura et al., 1963).

Philippines (Figure 2.11)
The Philippine archipelago is wedged between two opposing subduction zones. The South China Sea Basin, a backarc basin formed during Oligo-Miocene time (Taylor and Hayes, 1980), is being subducted along the Manila Trench in the west, and the West Philippine Basin of Eocene age (Mrozowski et al., 1982) along the Philippine Trench in the east. The west-facing arc comprises two segments, the Bataan arc and the Mindoro arc (DeBoer et al., 1980; Defant et al., 1988), which are divided by a major zone of volcanism trending NE-SW through southern central Luzon, the Macolod Corridor. Although no seismic zone

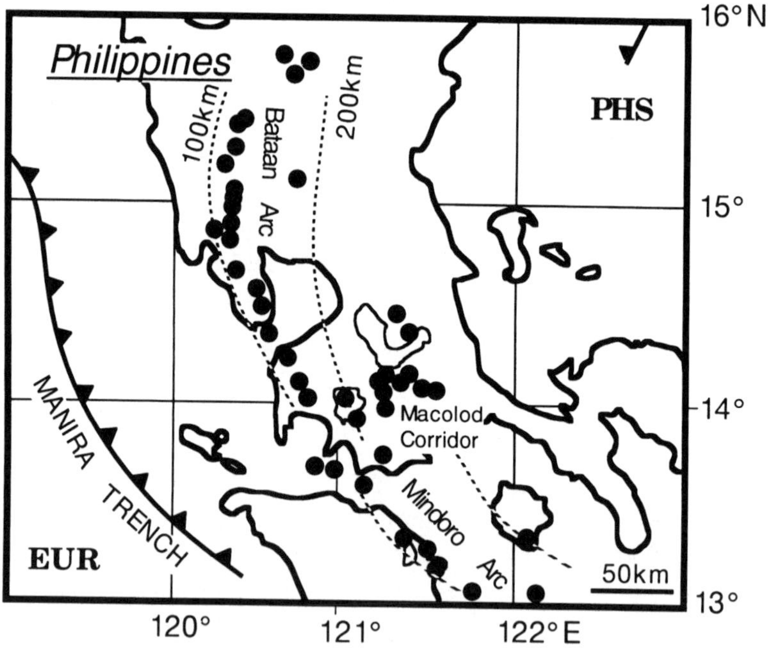

Figure 2.11 Distribution of volcanoes (solid circles) and depth contours to the surface of the dipping sesimic zone for the Bataan and Mindoro arcs, Philippines. These arcs are separated by the Macolod Corridor. PHS = Philippine Sea plate, EUR = Eurasian plate.

has been documented beneath the Macolod Corridor, dipping seismic zones have been traced to 200 km beneath the Bataan and Mindoro arcs (Cardwell et al., 1980; Hamburger et al., 1982; Defant et al., 1988). Two subparallel volcanic chains exist within the Bataan arc and are approximately 100–120 and 180–200 km above the Wadati-Benioff zone. The Mindoro arc may also comprise two volcanic chains. The spacing between individual volcanoes is noticeably closer among those nearer the trenches in these subduction zone systems.

Sangihe (Figure 2.12)

The Molucca Sea plate is subducting both westward beneath the Eurasian and eastward beneath the Philippine Sea plates to form the Sangihe and the Halmahera arcs in eastern Indonesia. This is part of a greater region of complex interaction between the Eurasian, Philippine Sea, Indo-Australian, and Pacific plates and numerous microplates (e.g., Hamilton, 1979). The Wadati-Benioff zone is well developed in the Sangihe arc, dipping at 55° to 65° and reaching a depth of about 650 km (Hatherton and Dickinson, 1969; Cardwell et al., 1980; McCaffrey, 1982). The Sangihe arc includes more than 25 Quaternary volcanoes, across a width of 50 km and extending about 500 km north from Sulawesi to southern Mindanao. Two subparallel volcanic chains are located 100–105 km and 150–180 km above the surface of the subducting plate.

Indonesia (Figure 2.13)

The Indonesian subduction zone system extends some 5000 km from Sumatra to the Banda Sea and has been subdivided into the Sumatra, Sunda, and Banda volcanic arcs (e.g., Hutchison, 1982). More than 80% of the volcanoes in Indonesia are concentrated in the Sunda arc. The Indo–Australian plate is subducted northward beneath the Eurasian plate and forms a dipping seismic zone down to about 600 km (Hamilton, 1979). Second or backarc-side volcanic chains are present in the western half of Java (105–110°E), Lombok Sumbawa (~117°E), and Flores (120°E–123°E), with depths to the surface of the subducting slab being 110–120 km and 200 km beneath the volcanic front and the second chains, respectively. An alkaline (high K to shoshonitic) volcano, Muriah, lies to the north of these main volcanic chains and about 400 km above the Wadati–Benioff zone (Hutchinson, 1982; Edwards et al., 1991).

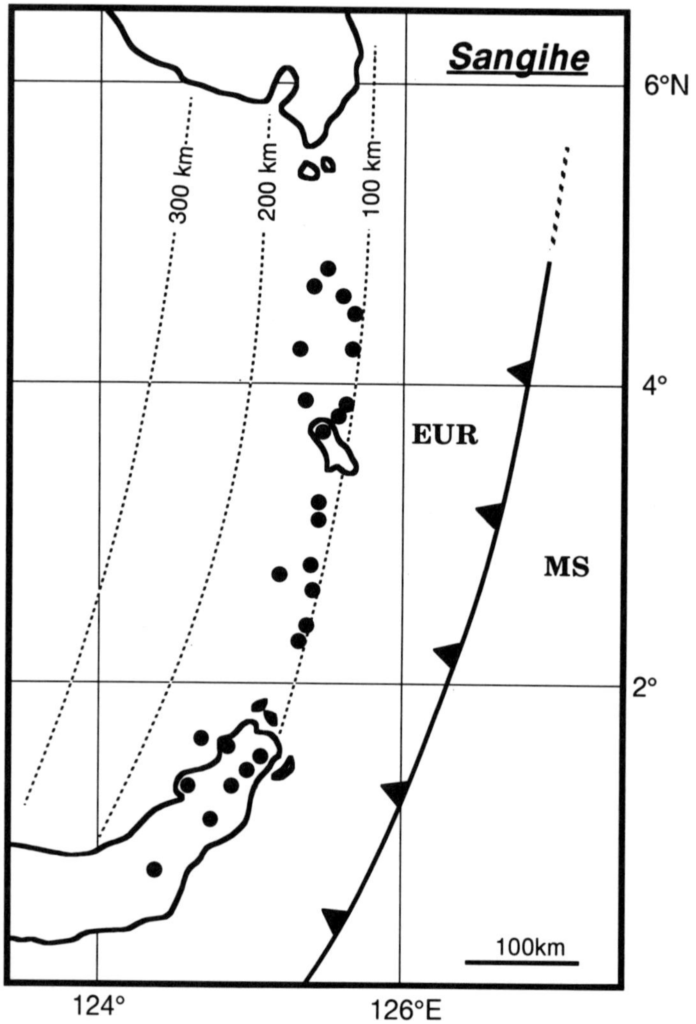

Figure 2.12 Distribution of volcanoes (solid circles) and depth contours (dashed lines) to the surface of the dipping sesimic zone for the Sangihe arc. MS = Molucca Sea plate, EUR = Eurasian plate.

North Island, New Zealand (Figure 2.14)

North Island volcanism is related to subduction of the Pacific plate at the Hikurangi Trench beneath the Indo–Australian plate. The seismic zone dips ~50° toward the northwest and can be traced to a depth of ~250 km (Ansell and Adams, 1986; Bannister, 1988). This subduction

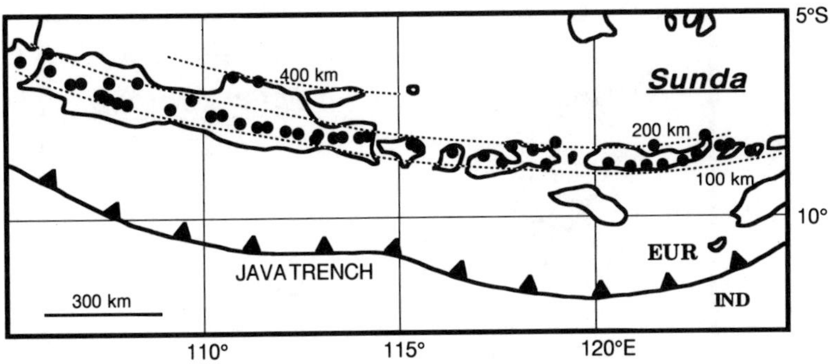

Figure 2.13 Distribution of volcanoes (solid circles) and depth contours (dashed lines) to the surface of the dipping sesimic zone for the Sunda arc. IND = Indo-Australian plate, EUR = Eurasian plate.

zone is characterized by extensive subduction-related volcanism and rifting along the volcanic arc. The Taupo volcanic zone (e.g., Cole, 1982) forms the volcanic front at 100–120 km above the top of the subducting lithosphere, and to the west is the Taupo-Rotorua Depression, a rift basin with a maximum width of 50 km (Cole, 1984; Stern, 1985). The occurrence of a backarc-side volcanic chain located ~200 km above the subducting slab has been proposed by Tatsumi and Tsunakawa (1992), and includes much smaller volcano numbers and volumes than the volcanic front. Further to the northwest, in the Northland Peninsula, alkalic to tholeiitic basalts crop out over an area of ~2500 km^2 (Weaver and Smith, 1989). These Northland basalts have trace element characteristics similar to intraplate, ocean island basalts (Heming, 1980; Weaver and Smith, 1989; Tatsumi and Tsunakawa, 1992), though the presence of weak Nb depletions may indicate the involvement of a subduction zone magma source during their genesis (Weaver and Smith, 1989).

Andes (Figures 2.15 and 2.16)

The Andes form an ~4000-km-long mountain chain parallel to the Peru-Chile Trench, where the Nazca and other east Pacific plates are subducted beneath the South American plate. Active volcanism is limited to regions where the seismic zone is steeply dipping, in a northern volcanic zone from southern Columbia to northern Ecuador, a central volcanic zone from southern Peru to northern Chile, and a southern volcanic zone in southern Chile (e.g., Thorpe et al., 1982).

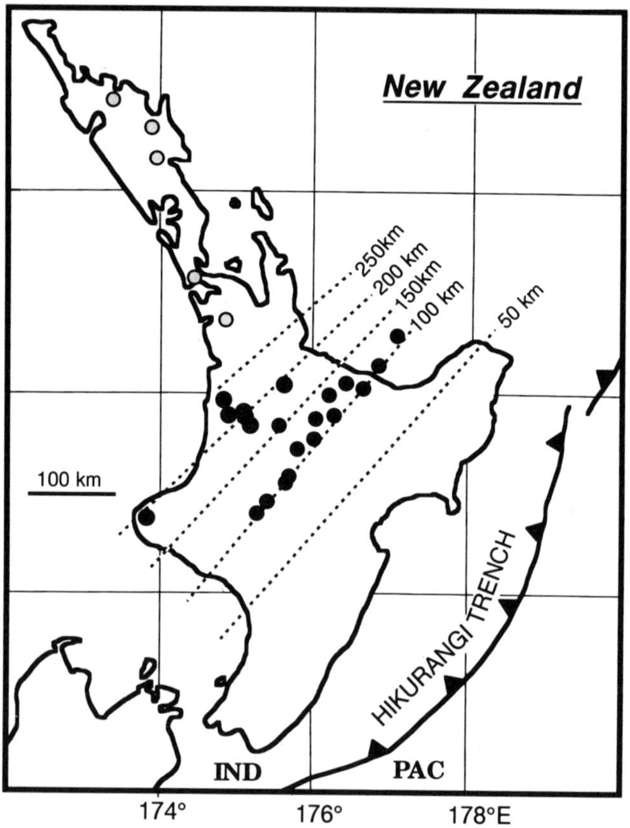

Figure 2.14 Distribution of volcanoes (solid circles) and depth contours (dashed lines) to the surface of the dipping sesimic zone for the North Island, New Zealand. "Intraplate" volcanism occurs in the backarc region (open circles). PAC = Pacific plate; IND = Indo-Australian plate.

These active volcanic zones are separated by regions beneath which very young oceanic lithosphere and active spreading ridges are subducted, at shallower-dipping angles (Barazangi and Isacks, 1979). Two volcanic chains are developed in each of the three volcanic zones and are located 100–120 km and 150–200 km above the surface of the subducting plates (Cahill and Isacks, 1992). Plateau lavas have been erupted periodically since the late Cretaceous or early Cenozoic in the Patagonian region behind the backarc-side volcanic chain in the southern volcanic zone (e.g., Stern et al., 1990). The Pliocene to Quaternary lavas of this province are largely alkaline basalt, whereas older lavas are subalkaline tholeiites.

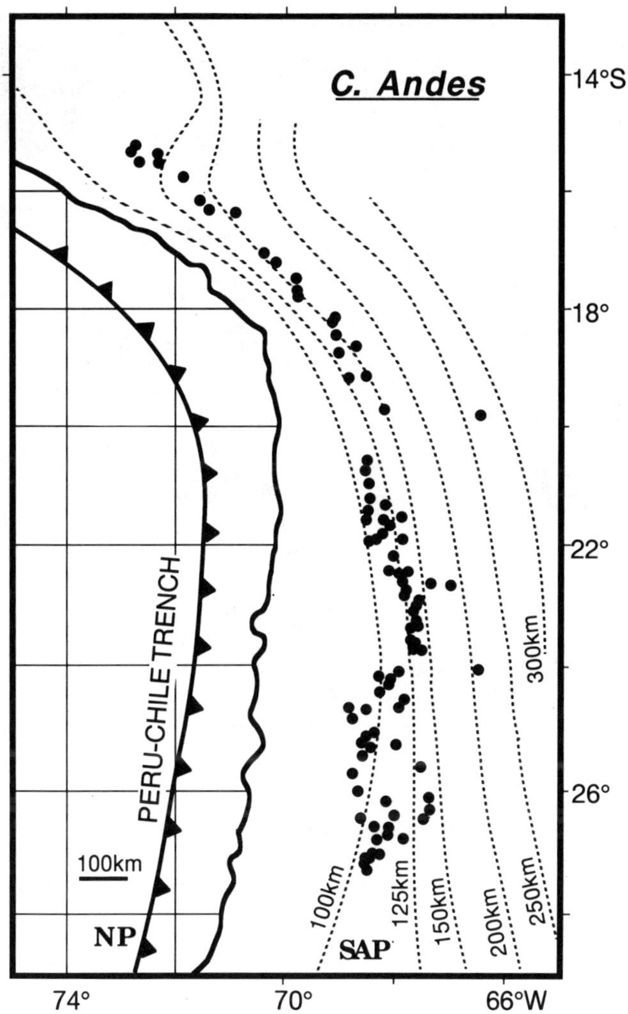

Figure 2.15 Distribution of volcanoes (solid circles) and depth contours (dashed lines) to the surface of the dipping sesimic zone for the central Andes. NP = Nazca plate, SAP = South American plate.

Central America Arc (Figure 2.17)

The Central America arc is composed of a single volcanic chain divided into eight segments (e.g., Carr et al., 1982; Carr, 1984) separated by transverse faults and offsets which correspond to changes in the strike and dip of the dipping seismic zone. The depth beneath the volcanic chain to the top of the subducted Cocos plate is constant within the individual segments, but varies between 100 and 150 km from seg-

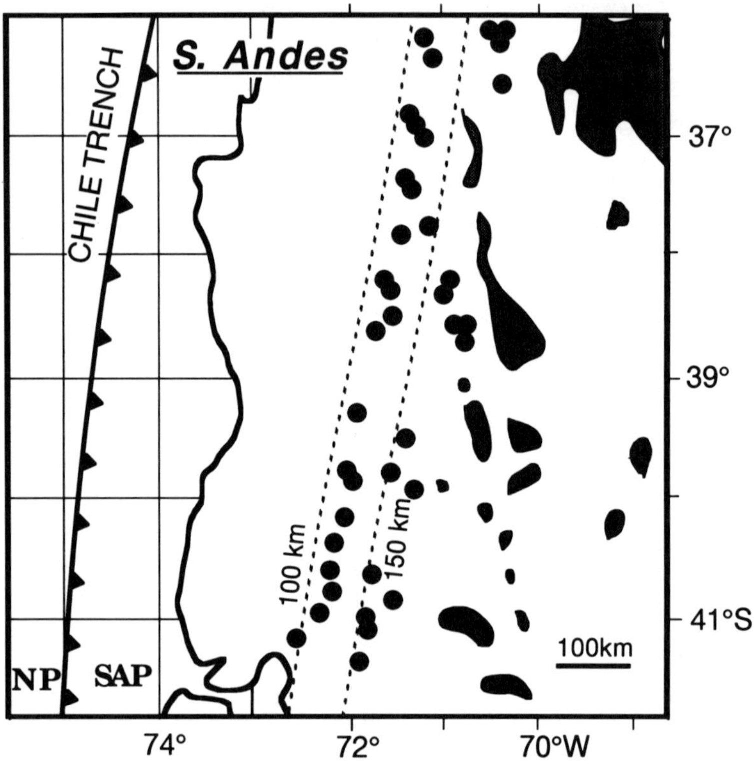

Figure 2.16 Distribution of volcanoes (solid circles) and depth contours (dashed lines) to the surface of the dipping sesimic zone for the southern Andes. The Patagonian plateau lavas are shown as hatched fields in the backarc region. NP = Nazca plate, SAP = South American plate.

ment to segment. The segmentation is also manifested in the spacing, size, and chemistry of volcanic centers, with segment boundaries possessing smaller and more widely spaced volcanic centers, and magmas in the center of each segment tending to be more enriched in incompatible elements at equivalent SiO_2 contents.

Cascade Arc (Figure 2.18)

The Cascade Range of western North America is sited above the subducting Juan de Fuca plate. This young subducting plate does not form a clearly defined dipping seismic zone at depth, having been traced to a depth of only 60 km beneath the northern and southern Cascade Range (Weaver and Baker, 1988). The location of the volcanic front is

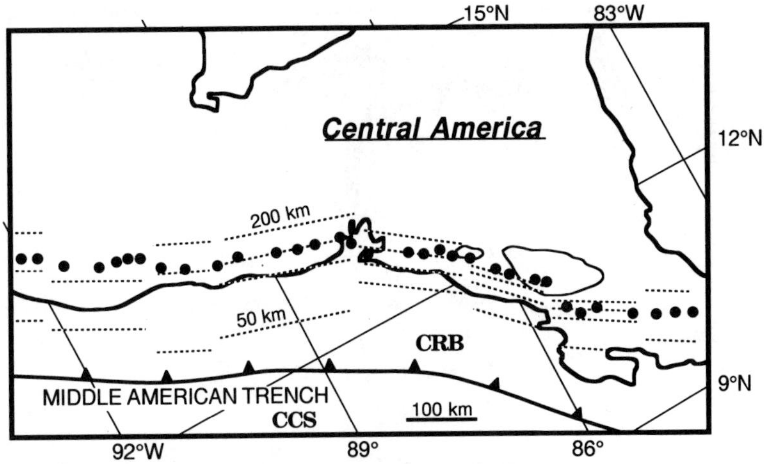

Figure 2.17 Distribution of volcanoes (solid circles) and depth contours (dashed lines) to the surface of the dipping sesimic zone for Central America. CCS = Cocos plate, CRB = Caribbean plate.

difficult to discern, as is evidence for two volcanic chains, especially at the northern and southern ends of the volcanic arc. Guffanti and Weaver (1988) have divided the Cascade Range into six segments, including one which contains the High Lava Plains, which occur along the northern margin of the Basin and Range province and lie perpendicular to the volcanic arc.

South Sandwich Arc (Figure 2.19)
The South Sandwich (or Scotia) arc is formed in a complex zone of microplates at the boundary of the major South American and Antarctic plates. Active backarc spreading has taken place for about 8 Ma behind the volcanic arc (e.g., Baker, 1978), forming the Scotia and Sandwich microplates on opposing sides of the spreading centers. The South Sandwich arc is over 500 km in length and comprises 11 volcanic islands (Baker, 1978), 10 of which are equally spaced by ~60 km along the volcanic front at ~100–120 km above the top of the subducting lithosphere (Forsyth, 1975), and a single volcano (Leskov) at a height ~160–180 km.

2.2.2 Number of volcanoes and volume of volcanic materials

A clear difference between trench- and backarc-side volcanic chains is the greater number of volcanoes and volume of magmatism which

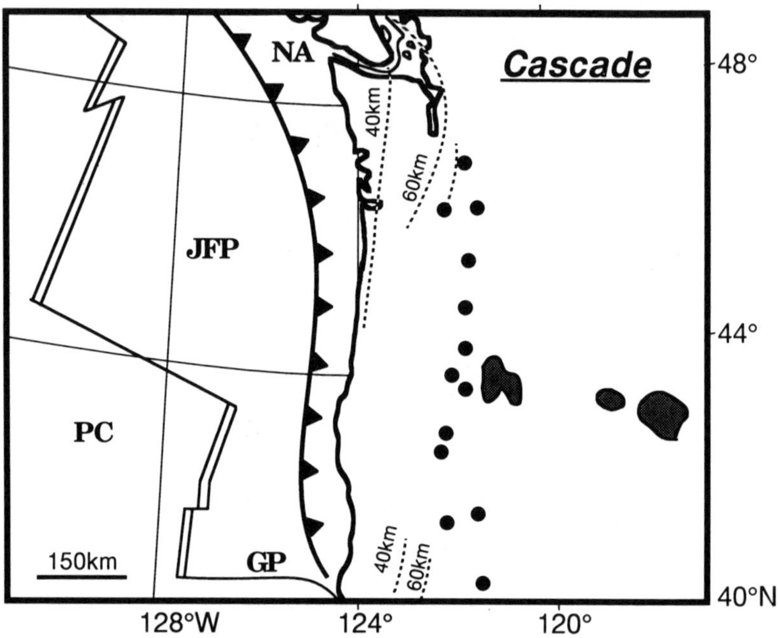

Figure 2.18 Distribution of volcanoes (solid circles) and depth contours (dashed lines) to the surface of the dipping sesimic zone for the Cascade region. No seismic slab is documented beneath volcanoes. The High Lava Plains, a volcanic chain perpendicular to the volcanic arc, is located in the backarc region. JFP = Juan de Fuca plate, GP = Gorda plate, PC = Pacific plate.

typify the trench-side chains (Figure 2.7). In the Quaternary NE Japan arc, the number of volcanoes is not distributed bimodally across the arc (Figure 2.7), but the volume of volcanism is, and a far greater volume of volcanic materials has been erupted along the trench-side volcanic chain (Figure 2.7). These observations were first made by Sugimura et al. (1963) and reconfirmed by Aramaki and Ui (1982). The volume difference is magnified further if the volume of pyroclastic flows and fall deposits associated with large calderas are taken into account, since large calderas are far more common along the trench-side volcanic chain. Similar volume differences also characterize other arc systems, such as the Central Andes, where the volume of large late Neogene felsic systems is ~25,000 km^3 and 5000 km^3 in the trench- and backarc-side, respectively (de Silva and Francis, 1991).

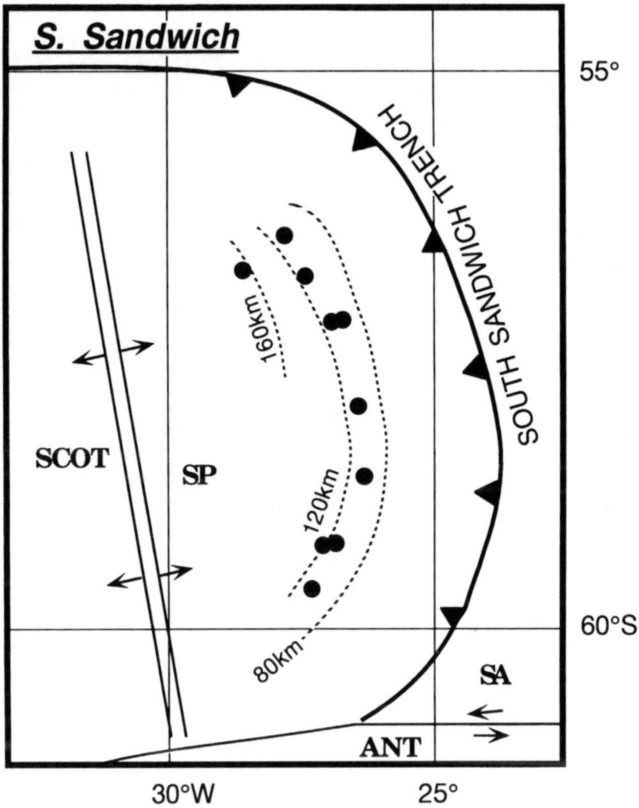

Figure 2.19 Distribution of volcanoes (solid circles) and depth contours (dashed lines) to the surface of the dipping seismic zone for the South Sandwich arc. SCOT = Scotia plate, SP = Sandwich plate, SA = South America plate, ANT = Antarctic plate.

The average volumes of individual volcanoes in the NE Japan arc have been estimated by Aramaki and Ui (1982) and shown to be similar in both the trench- and backarc-side volcanic chains (20 and 28 km^3, respectively). Moreover, minimum FeO*/MgO ratios of lavas tend to be smaller for trench-side than backarc-side volcanoes in NE Japan (Kushiro, 1983), a feature also observed in New Zealand and documented for the Kurile arc, where FeO*/MgO ratios average 2.13 versus 1.72, 1.65 versus 1.03, and 2.14 versus 1.61, for trench- and backarc-side volcanoes in the north, middle, and south Kuriles, respectively (Avdeiko et al., 1991). This observation implies comparatively higher extents of differentiation for trench-side volcanoes and, taken together with the greater number and/or volume of trench-side volcanoes, requires that a much greater volume of magma be produced beneath the trench-side volcanic chain than the backarc-side chain. The origin of these differences will be evaluated in Chapter 4.

2.2.3 Volcanic arc width versus angle of subduction

The width of a volcanic arc, or the distance between trench- and backarc-side volcanic chains where present, varies appreciably, from almost 0 km in the Mariana arc to about 200 km in the Andes. Marsh (1979) noted that the width of a volcanic arc is inversely proportional to the angle of subduction, with wider volcanic arcs forming above shallower-dipping subduction zones. This is clearly illustrated by the Izu-Bonin-Mariana arc, where the width of the volcanic arc decreases as the dip angle of the downgoing Pacific plate changes from 30° beneath the Izu Peninsula in the north to nearly vertical in the Mariana region (Figure 2.20). This inverse correlation is observed widely as is a concurrent decrease in the width of the forearc, or trench-to-volcanic-front separation (Figure 2.21). It is notable that some volcanic arcs constitute only a single volcanic chain (i.e., have near zero width), for example, the Mariana, Tonga-Kermadec, Central America, and Lesser Antilles arcs, and that these arcs have steep angles of subduction (usually >60°). The only exception is the Tonga-Kermadec arc (~50°), which is complicated by the close spatial occurrence of active backarc spreading in the Lau Basin. It follows from these observations that the existence of two volcanic chains may not be discernible in steeply dipping subduction zones due to their effective overlap.

The negative correlation between the width of a volcanic arc and the angle of subduction provides compelling evidence for pressure-dependent controls on the production of magma in subduction zones. This pressure dependency can be emphasized further given the constant depths to the top of the subducted slab beneath the two volcanic chains (Figure 2.22). Trench-side volcanic chains overlie the surface of the subducting slab at a constant depth of 108 ± 18 km (1σ), the only exceptions being the Vanuatu arc (~150 km; Chiu et al., 1985) and the

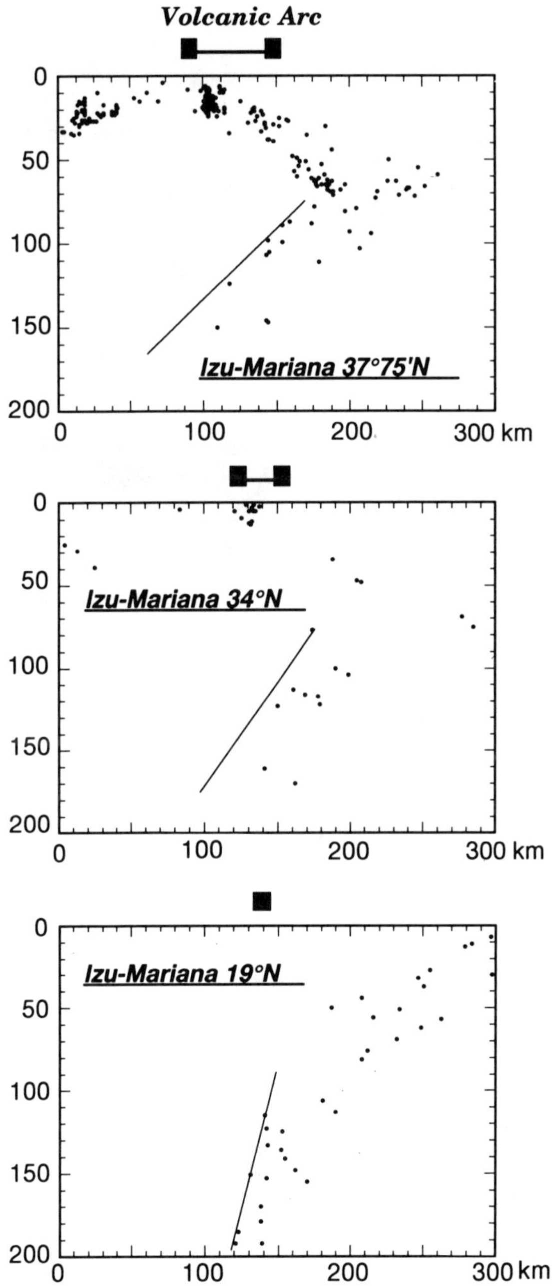

Figure 2.20 Cross sections through the Izu-Mariana arc, showing steepening of the dipping seismic zone and concomitant narrowing of the volcanic arc toward the south.

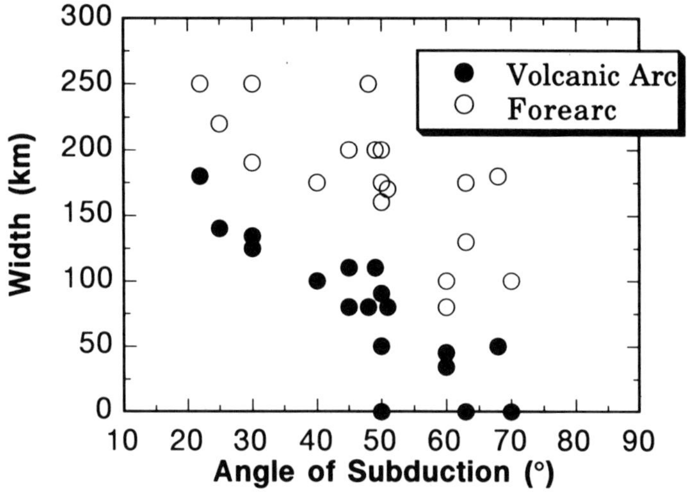

Figure 2.21 Negative correlation between the angle of subduction and the widths of the volcanic arc and forearc.

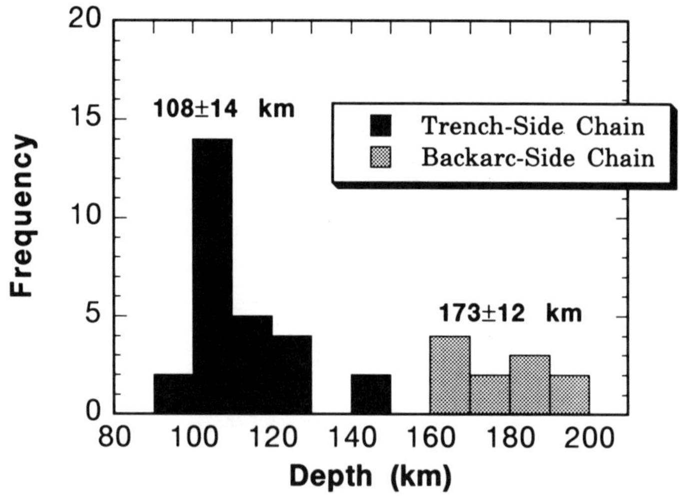

Figure 2.22 Histogram of depths to the surface of the subducted lithosphere beneath trench-side and backarc-side volcanic chains in subduction zones. Quoted values are the average depths and 1 standard deviation.

Eolian arc (~150 km; Mitropoulos and Tarney, 1992). Depths to the surface of the subducting lithosphere beneath the backarc-side chains are not as well constrained, due in part to sparser seismic observations; however, available data indicate a surprisingly constant average depth to the top of the subducting plate of 173 ± 12 km (Figure 2.22).

Explanations for the constancy in depth to the top of the subducting slab beneath the volcanic front have been proposed by many authors. The mechanisms responsible can be divided into two broad groups: (1) magmas able to reach the surface only behind the volcanic front, or (2) magma production in the mantle wedge being limited to behind the volcanic front. These opposing views are examined in detail in Chapter 4, where we provide evidence favoring the onset of melting beneath the volcanic front and formation of the trench-side and backarc-side volcanic chains as a consequence of pressure-dependent dehydration reactions in down-dragged hydrous peridotites at the base of the mantle wedge.

2.2.4 Examples of unusual volcano distributions

Examples of volcano distributions which are inconsistent with the spatial distribution systematics we have described are observed in a number of subduction zones. These cases represent significant departures from the development of dual, trench-parallel, volcanic chains, located at ~110 km and ~170 km above the top of the subducted lithosphere.

Unusual near-trench or forearc volcanism has been recognized in a number of subduction zone systems. The most extensive examples are of Eocene age and occur along much of the Izu-Bonin-Mariana forearc region of the western Pacific (e.g., Kuroda et al., 1978; Bloomer, 1987). More recently active examples have been described from North Tonga (Falloon et al., 1989), Vanuatu (Sigurdson et al., 1983), and Baja California (Saunders et al., 1987). The Setouchi volcanics occur in a similar forearc position in the SW Japan arc (Figure 2.23). Their ages (11–14 Ma; Tatsumi, 1983) preclude estimation of the depth to the top of the subducting lithosphere at the time of eruption; however, it seems probable that even shallower subduction of the younger, hence, hotter Philippine Sea plate that was formed at 25-17Ma took place during the Miocene.

These examples of near-trench volcanism are characterized by the

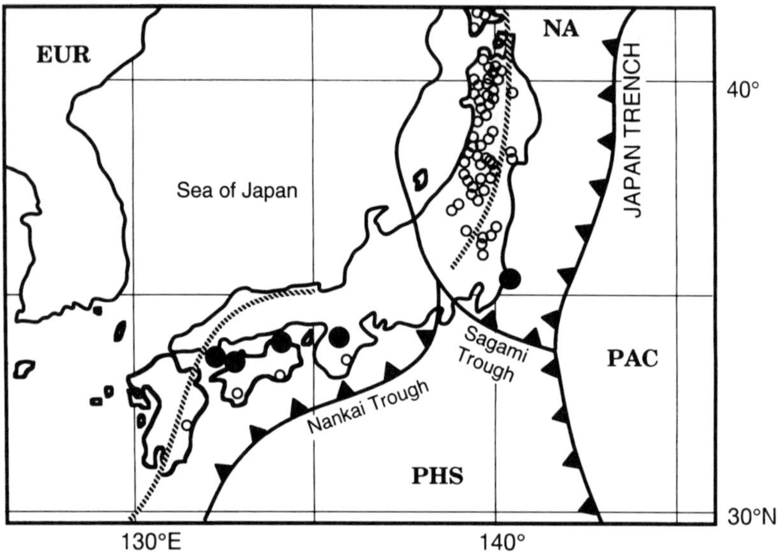

Figure 2.23 Distribution of high-Mg andesites (closed circles) and other Miocene volcanics (open circles) in the Japanese Islands. Quaternary volcanic fronts are shown as hatched lines, and reveal that Miocene volcanism took place much nearer the trench axis.

occurrence of suites of high-Mg andesite magmas (see Chapter 3). The Setouchi volcanics and other high-Mg andesites have been demonstrated to result from relatively hydrous partial melting of upper mantle peridotite at shallow levels (e.g., Kushiro and Sato, 1978; Tatsumi and Ishizaka, 1981). The presence of hot mantle beneath the forearc mantle wedge is widely considered necessary for the occurrence of near-trench volcanism and the formation high-Mg andesites. In this context, it is informative that in the Setouchi volcanics, and in North Tonga and Vanuatu, occurrences of high-Mg andesites are closely related, in both time and space, to active backarc basin spreading. Tatsumi and Maruyama (1989) have argued, in the case of the Setouchi volcanics and similar near-trench volcanism at ~22 Ma in the NE Japan arc (Figure 2.23), that cold forearc mantle wedge materials have been replaced by hotter asthenospheric materials at the same time as opening of the Japan Sea backarc basin was taking place during the Miocene. Interestingly, high-Mg andesites are often found in ophiolite belts and, as such, hold important palaeotectonic implications for the origin of these ophiolites that require formation in a forearc subduction zone setting followed by later obduction onto continental margins.

The occurrence of volcanism in the far backarc regions of the Kamchatka (Figure 2.9), Chile (Figure 2.16), and Cascade (Figure 2.18) arcs presents another unusual case. These examples take the form of lava fields rather than the discrete central volcanoes which typify most subduction zone volcanism. They also occupy unusual tectonic settings: the subduction of oceanic ridges or young hot lithosphere in the Chile and Cascade cases, and the occurrence of a transform-subduction plate boundary in the Kamchatka case. Tatsumi and colleagues (1994a) have speculated on this basis that atypically high temperatures may occur in such tectonic settings and may be responsible for partial melting of mantle wedge materials at greater depths than is normally the case. In Kamchatka, for example, modeling of the thermal structure surrounding the edge of the subducting plate reveals that temperatures upon the upper surface at the subducting plate's edge are 200–300°C higher than the plate temperature ~50 km from the edge. Thus, it may be entertained that partial melting of either the subducting plate or of the base of the mantle wedge under atypically hot conditions may produce these examples of unusual magmatism.

A final unusual example of volcanism is the reversal in the typical across-arc distribution of volcano numbers and volume of volcanic material which occurred in the Miocene NE Japan arc (Figure 2.23). The increased volcano numbers and volumes toward the backarc in this case were associated with opening of the Japan Sea, and may reflect an accompanying change to the thermal structure of the mantle wedge and melt production therein (Tatsumi et al., 1989).

2.2.5 Volcano spacing

The occurrence of a characteristic regular spacing between volcanoes along volcanic arcs has been proposed in a number of cases, for example, a 70–75 km separation of volcanoes along the Central America (Carr et al., 1979), Aleutian, Alaska Peninsula, and Cascade arcs (Marsh and Carmichael, 1974). This characteristic spacing has been interpreted in terms of Rayleigh-Taylor instability wavelengths, formed where a less dense layer of partial melt is overlain by the subsolidus mantle wedge (Fedotov, 1975; Marsh, 1975, 1979). A negative correlation between volcano separation distance and the angle of subduction has also been noted (Marsh, 1979; Shimozuru and Kubo,

1983) and has been attributed to a decrease in compressional stress with increasing dip of subduction, due to reduced coupling between the subducting and overriding plates (Shimozuru and Kubo, 1983). However, though the average volcano spacing values might be similar from arc to arc (i.e., 58 ± 24; Shimozuru and Kubo, 1983), the spacing of volcanoes within individual arcs is often highly variable (Aleutian, 60 ± 33 km; Kamchatka, 39 ± 27 km; Kurile, 40 ± 30 km; NE Japan, 52 ± 32 km; Izu–Mariana, 86 ± 50 km; New Britain, 72 ± 72 km; Vanuatu, 93 ± 69 km; Solomon Islands, 100 ± 76 km; Tonga, 65 ± 39 km; Central America, 32 ± 23 km). It therefore remains to be demonstrated that volcano spacings are sufficiently regular to be of significance, an uncertainty which has been emphasized by Gill (1981), who dismissed volcano spacing as adding "nothing to genetic questions."

Although regular along-arc volcano separation is difficult to justify, a possible correlation between volcano numbers and the rate of subduction is worth examining. Acharya (1981) suggested that the number of volcanic eruptions is linearly related to the aseismic slip component of plate motion—the component of plate underthrusting that does not occur during large earthquakes—and further speculated that rare large-thrust earthquakes may cause significant shear heating along the surface of the subducting lithosphere and generate more

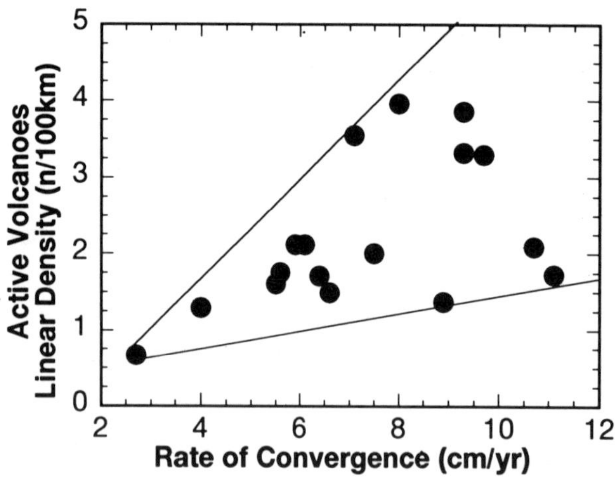

Figure 2.24 Correlation between the linear density of volcanoes in volcanic arcs and rate of convergence of the subducting and overriding plates. *(Data reproduced by permission from Shimozuru and Kubo, 1983.)*

magma as a consequence. However, shear heating is unlikely to be sufficient to cause magma generation, and it is not apparent how the rate of magma production in the mantle wedge can be related to the relatively shallow, crustal-level shearing events. On the other hand, a broad positive correlation is observed between the number density of arc volcanoes and the rate of plate convergence (Figure 2.24). This observation suggests that faster subduction results in greater melt production in the mantle wedge. Given that the release of hydrous fluid from the subducting oceanic crust drives magma generation, then a greater fluid flux deriving from dehydration of larger volumes of subducting oceanic lithosphere is likely to enhance melt generation and may ultimately be linked to larger volcano numbers through increased formation rates of gravitational instabilities within the magma source region (this will be discussed in more detail in Chapter 4).

3 Geochemical Characteristics of Subduction Zone Magmas

3.1 General Geochemical Characteristics

Subduction zone magmas are noted for their distinct chemistry compared with magmas in the other tectonic settings. In particular, they possess relative enrichments and depletions in certain elements and element groups compared to mid-ocean-ridge basalts (MORB) and ocean island basalts (OIB) (Figure 3.1). Considerable attention has been focused on the large-ion-lithophile-element (e.g., Cs, Rb, K, Ba, Sr) over-abundances and high-field-strength-element (HFSE) (e.g., Nb, Zr, Ti) depletions, since the formation of these particular chemical fractionations is likely to provide key insights into the processes responsible for generating subduction zone magmatism.

Subduction zone magmas are also notable for their compositional diversity and generally more evolved bulk compositions compared with other tectonic settings. These features of subduction zone magmas are partly due to protracted crustal-level differentiation processes, as in the classical evolution of basalt to basaltic andesite, andesite, dacite, and rhyolite. However, chemical diversity also arises from the existence of a broad spectrum of primary magma compositions, which can be attributed to variation in the pressure and temperature of melting, volatile species and their abundances, source composition, and so on.

3.1.1 Classification

The compositional diversity of subduction zone magmas necessitates the adoption of a classification scheme so that the distinction and comparison of different magma types and their petrogeneses can be

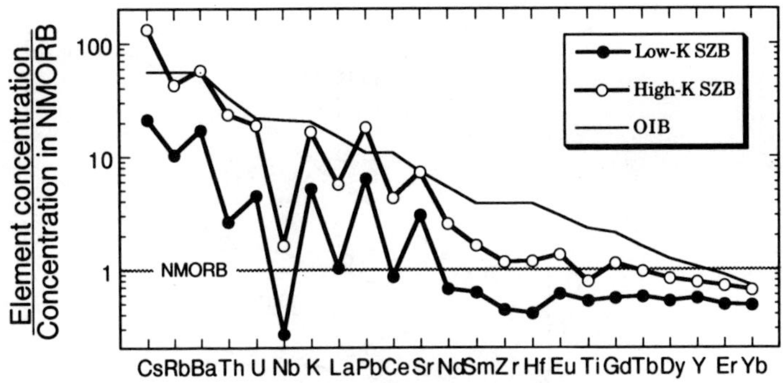

Figure 3.1 Multielement diagram comparing "average" low-K and high-K subduction zone basalt (SZB) compositions with the average compositions of normal mid-ocean-ridge basalt (MORB) and oceanic island basalt (OIB) compiled by Sun and McDonough (1989). All element concentrations are normalized to the values of average N-MORB. *(Reproduced by permission from Sun and McDonough, 1989.)*

made. Numerous classification schemes have been applied to subduction zone magmas, but their use has been plagued by inconsistencies. Considerable confusion currently exists, particularly with regard to the meaning and significance of nomenclature such as calc-alkaline, tholeiitic, and high-alumina basalt, among others. To circumvent this confusion, we recommend use of a simple classification scheme based on K_2O versus SiO_2 variation, and, for clarity, we review the use of certain rock names.

A Recommended Classification Scheme

Subduction zone magmas can be classified by using chemical, mineralogical, or both criteria. We believe it is most practical and simple to subdivide subduction zone magmas based on their K_2O and SiO_2 contents, as in Figure 3.2. This classification scheme identifies shoshonitic, high-K, medium-K, and low-K magma series and distinguishes basalt, basaltic andesite, andesite, dacite, and rhyolite compositions at SiO_2 contents of 48 wt%, 53 wt%, 57 wt%, 63 wt%, and 70 wt%. Examples of K_2O versus SiO_2 variation trends for selected subduction zone magmatic suites are illustrated in Figure 3.3, which shows that most suites evolve subparallel to the K_2O discriminant lines and so correspond to a single magma series. This scheme is very similar to the TAS (total alkalis versus silica) scheme commonly applied to igneous rocks (LeBas et al., 1986) but is preferrable in this case because the

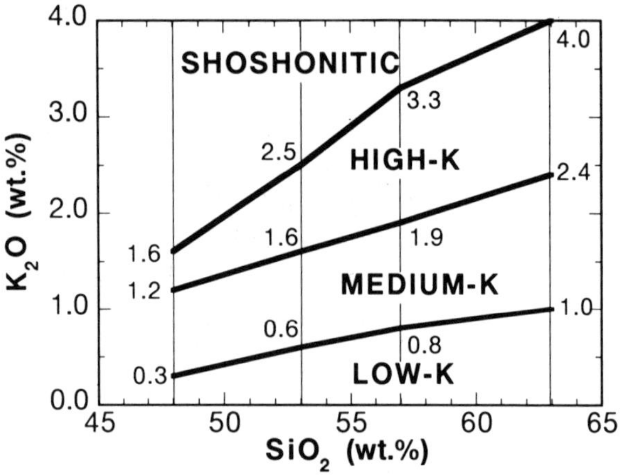

Figure 3.2 Recommended K_2O versus SiO_2 classification scheme for subduction zone magmas. SiO_2 contents of 53 wt%, 57 wt%, and 63 wt% separate basalt, basaltic andesite, andesite, and dacite compositions.

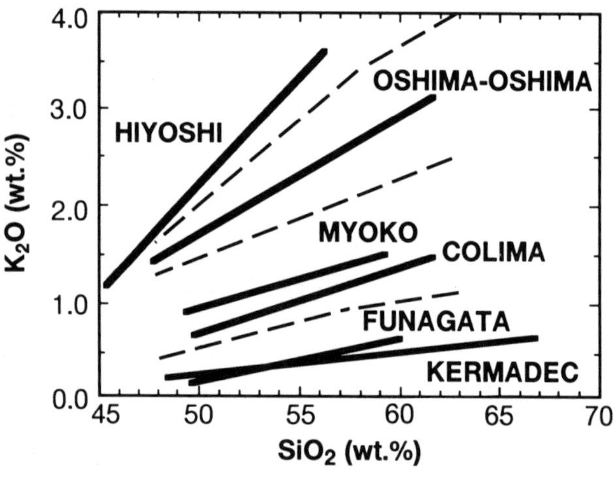

Figure 3.3 Typical differentiation trends for volcanoes within the range of magma series compositions observed in subduction zones. Dashed lines are the magma series boundaries shown in Figure 3.2.

TAS scheme does not adequately differentiate the broad spectrum of subduction zone magma compositions nor take into account potassic magma types (e.g., shoshonites).

An important class of subduction zone magmas that is not incorporated in the K_2O versus SiO_2 classification scheme is the high-magnesian andesites and their associated magma series. High-magnesian andesites, also referred to as boninites, are characterized by relatively high MgO and high-SiO_2 compositions, and are defined as having >53 wt% SiO_2 and Mg# $(100 \times Mg/(Mg + Fe)) > 60$. Members of the related magma series need to be demonstrably derived from parental magmas falling within these compositional bounds (Crawford et al., 1989).

Five broad types of subduction zone primary magmas can be recognized within the foregoing classification framework, namely high-Mg andesite and shoshonitic high-, medium-, and low-K series primary magma compositions. It should be remembered, however, that these are functional definitions only and that a continuous spectrum of primary magma compositions probably exists between these five types.

Additional Nomenclature: Alkaline versus Subalkaline

The division of igneous rocks into alkaline and subalkaline types serves as a fundamental distinction of magma type in igneous petrology and geochemistry. This division is related to a magma's silica activity and can be defined either chemically or mineralogically on the basis of reactions such as

$$\underset{\text{orthopyroxene}}{2(Mg,Fe)SiO_3} \rightarrow \underset{\text{olivine}}{(Mg,Fe)_2SiO_4} + SiO_2$$

and

$$\underset{\text{albite}}{NaAlSi_3O_8} \rightarrow \underset{\text{nepheline}}{NaAlSiO_4} + 2SiO_2$$

This is typically performed by recasting bulk-rock chemical compositions into hypothetical rock-forming (normative) minerals and classifying according to the calculated normative mineral assemblage (Figure 3.4). Alkaline magmas have normative olivine and feldspathoids (e.g., nepheline), whereas subalkaline magmas have normative orthopyroxene and plagioclase and can be further divided into olivine and quartz normative types (Figure 3.4). A strength of this classification scheme is that the alkaline and subalkaline magma series tend to remain separated during differentiation since early crystallization of olivine, clinopyroxene, and plagioclase from basaltic magmas drives derivative magmas farther from the "critical plane of silica under saturation" which divides the magma series (Figure 3.4).

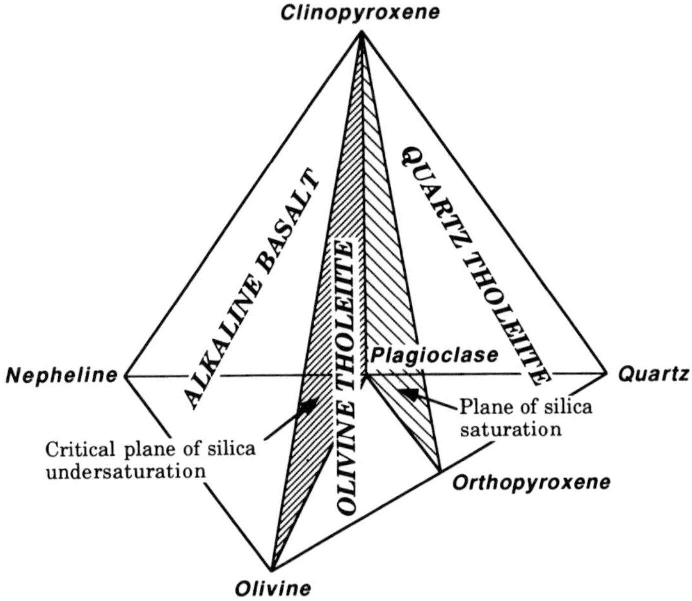

Figure 3.4 Classification of basalts based on their normative mineralogy in the basalt tetrahedron. Quartz tholeiites have normative quartz, olivine tholeiites have normative hypersthene and olivine, and alkali basalts have normative olivine and nepheline. *(Reproduced by permission from Yoder and Tilley, 1962.)*

Mineralogical criteria can be applied on a similar basis to distinguish alkaline and subalkaline magmas. Tilley (1950) and Kuno (1959, 1960) distinguished these magma types based on the presence or absence, respectively, of a reaction relation between olivine and liquid producing a rim of Ca-poor pyroxene (e.g., orthopyroxene) around groundmass olivine. The absence of the reaction relationship has become an accepted means of recognizing alkaline rocks, although it strictly applies only to highly-silica-undersaturated alkaline rocks.

Subdivision of Subalkaline Magmas: Calc-Alkaline and Tholeiitic

Tholeiitic and calc-alkaline magma types were first distinguished within the subalkaline series during the 1930s (e.g., Peacock, 1931; Kennedy, 1933). Kennedy (1933) proposed the existence of two different primary magma series of olivine-basalt magma type and tholeiitic magma type, with olivine-basalts containing olivine, augite, plagioclase, and magnetite phenocrysts and differentiating to silica-undersaturated alkaline rocks, whereas tholeiites contained pyroxene, plagioclase, and magnetite phenocrysts and differentiated to silica-saturated and -oversaturated rocks. Peacock's classification divided igneous rocks into alkaline, alkali-calcic, calc-alkaline, and calcic rock series, based on the "alkali-lime index" (SiO_2 content at which $Na_2O + K_2O = CaO$), but has since become largely redundant. Nockolds and Allen (1953, 1954, 1956) proposed the use of tholeiitic and calc-alkaline for magma series with and without iron enrichment with increasing SiO_2 contents, respectively. The calc-alkaline and tholeiitic series were subsequently defined quantitatively by Miyashiro (1974), who proposed that calc-alkaline and tholeiitic rock series show steeper and gentler slopes, respectively, than the straight line

$$SiO_2\ (\text{wt\%}) = 6.4 \times FeO^*/MgO + 42.8$$

Miyashiro (1974) further suggested that calc-alkaline and tholeiitic series could be distinguished as lying above and below the line, in cases where magmas have undergone intermediate degrees of differentiation (i.e., have $2.0 < FeO^*/MgO < 5.0$). However, Miyashiro's discrimant line is frequently misused and applied as a simple compositional discriminant outside this compositional range rather than as a "trend slope" comparison for which it was primarily intended.

The inconsistent application of the name calc-alkaline, outside the foregoing definitions, has ranged from a general term for all subduction zone magmas (e.g., the Basaltic Volcanism Study Project, 1981) to that of a magma series intermediate between low-K and high-K (alkaline) series magmas (e.g., Wilson, 1989). While it is true that calc-alkaline andesites are generally more enriched in alkali elements than tholeiitic andesites within a particular region, low-K andesites without iron-enrichment, that is low-K calc-alkaline andesites, *do* exist (Gill, 1981). Moreover, medium-K magmas that conform to a tholeiitic trend also occur (Figure 3.5). The interchangeable use of the terms *medium-K* and *calc-alkaline* series is therefore discouraged. The calc-alkaline series is best defined by the absence of iron enrichment in subalkaline rocks and, in practice, applied using Miyashiro's definition.

High-Alumina Basalts

High-alumina basalts were originally defined as aphyric basalts with Al_2O_3 contents greater than 17 wt%. They were first documented in a

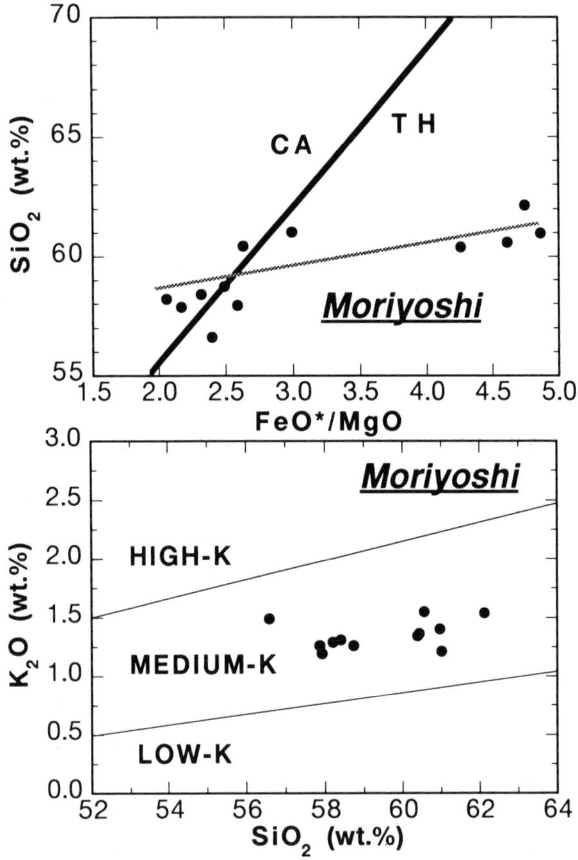

Figure 3.5 Compositional variation of lavas from Moriyoshi volcano, NE Japan. These rocks form a tholeiitic (TH) trend yet belong to the medium-K series, which is often associated with calc-alkaline (CA) series *(Reproduced by permission from Nakagawa, 1983.)*

subduction zone setting from Medicine Lake Highland by Powers (1932). Kuno (1960, 1966) observed that high-alumina basalts in Japan possess intermediate characteristics between tholeiitic and alkaline basalts in terms of distribution, alkali contents, and mineralogical characteristics. This prompted Kuno (1966) to propose a total alkali versus SiO_2 diagram, which separated alkaline, high-alumina basalts, and tholeiitic basalts. It has since become accepted that primary high-alumina basalts can be produced at depths intermediate between tholeiitic and alkaline magmas (e.g., Green et al., 1967; Tatsumi et al.,

1983), though alternative origins have also been proposed (Chapter 5).

Recently, the term *high-alumina basalt* has been applied less rigorously to include porphyritic magmas, in particular to plagioclase phenocryst-rich basalts occurring in subduction zones. Brophy and Marsh (1986) and Brophy (1987), among others, have argued that these particular magmas are the product of melting of the subducting oceanic crust; however, Crawford et al. (1987) have since demonstrated that they originate by crystal-liquid differentiation and plagioclase accumulation from more magnesian parent basalts.

3.1.2 Phenocryst assemblages

A feature of subduction zone volcanic rocks is their frequently crystal-rich or highly porphyritic nature (Figure 3.6). Ewart (1976) has shown that subduction zone magmas tend to display bimodal distributions in total phenocryst contents (aphyric and ~30%; Figure 3.6), and suggested this results from fractional crystallization and/or flow differentiation within a magma reservoir or a conduit.

Phenocryst assemblages and the sequence of phase crystallization in subduction zone magmas display certain systematic variations with bulk composition and with magma series or type. Figure 3.7 shows schematically the crystallization sequences observed in three of the principal magma series as a function of temperature (or magma differentiation). The temperature used in this figure is based on the liquidus temperatures of estimated primary basalt compositions (Tatsumi et al., 1983) and the crystallization temperature of hornblende and biotite in hydrous andesite compositions (e.g., Wyllie, 1977). The following two points are notable: (1) The stability fields and liquidus temperatures of plagioclase and quartz, which are strongly affected by the H_2O content of a magma, are smaller and lower, respectively, in the high-K magma series. (2) Orthopyroxene is the second ferromagnesian liquidus mineral to crystallize in some NE Japan arc magma suites, whereas clinopyroxene is more common in most other subduction zone settings. The former suggests higher H_2O contents in more high-K magma series, whereas the latter may reflect higher degrees of partial melting for the NE Japan orthopyroxene-olivine basalts, given normative diopside (clinopyroxene) and normative diopside/hypersthene ratios are known to decrease with increasing degrees of partial melting of depleted, clinopyroxene-free peridotite (Kushiro, 1969).

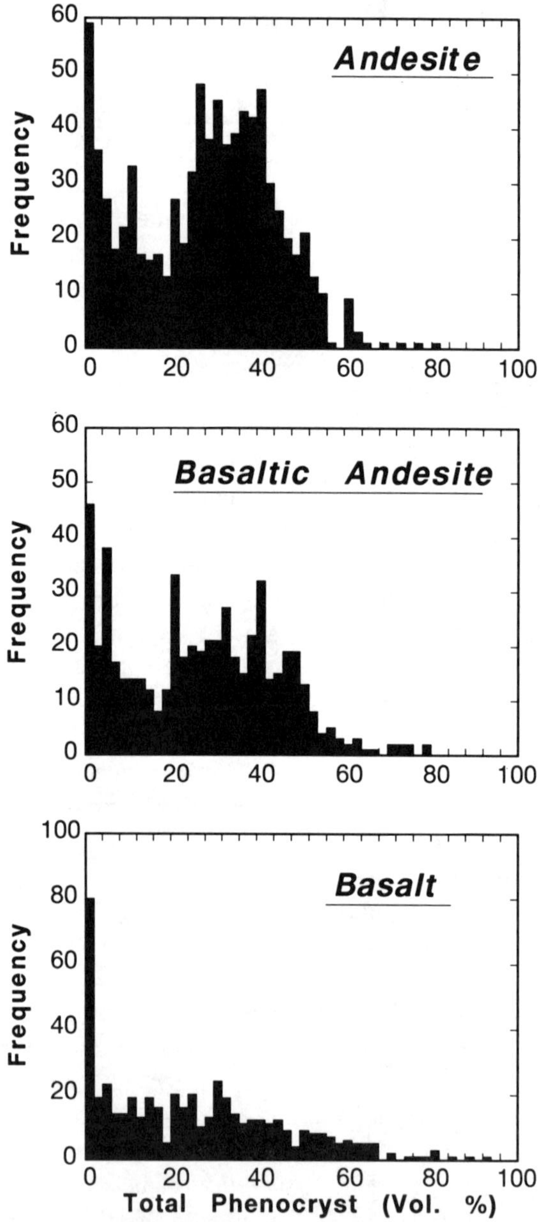

Figure 3.6 Frequency distributions for total phenocryst contents in compilations of subduction zone andesites, basaltic andesites, and basalts. Note the dominance of highly porphyritic magmas and the bimodal populations.

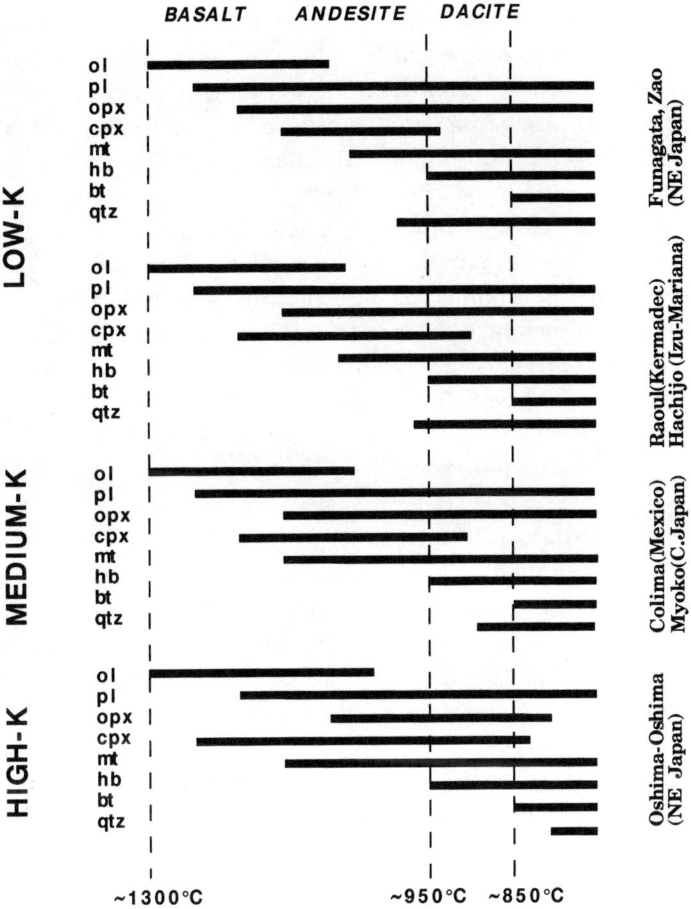

Figure 3.7 Schematic crystallization sequences for low-, medium-, and high-K series magmas in subduction zones. Temperatures are based on the experimentally determined liquidus temperatures of olivine in primitive subduction zone magmas and of hornblende and biotite in andesitic and dacitic compositions (Wyllie, 1977; Tatsumi et al., 1983). The major difference in these sequences occurs with the order of plagioclase and pyroxene crystallization and with the order of quartz, hornblende and biotite crystallization. Phenocryst abbreviations are ol, olivine; pl, plagioclase; opx, orthopyroxene; cpx, clinopyroxene; mt, magnetite; hb, hornblende; bt, biotite; qtz, quartz.

3.1.3 Subduction zone primary magmas

Almost all magmas erupted on the Earth's surface have had their original compositions modified by shallower-level processes such as fractional crystallization, magma mixing, and/or crustal contamination. These differentiation processes mask both the original (primary) magma composition and the chemical signature of the deeper processes which have contributed to form the magma. Accordingly, if the origin of magmatism is to be assessed it is necessary to establish the nature of the most primitive magmas from which other magmas in a suite or series have evolved. Failure do this can (and has) lead to erroneous conclusions about magma generation processes. An example is the primary status once widely attributed to the voluminous andesites that dominate some subduction zones and which were interpreted as melts generated from the subducting ocean crust. This view has now faded as new studies have brought to light increasing evidence for the evolution of andesites and other evolved magmas from primitive, mantle-derived, basaltic parents via shallow, crustal-level differentiation processes.

Differentiation of Primitive Magmas

Given our intention to evaluate the primitive chemical characteristics of subduction zone magmas so that the hidden processes which occur at depth and give rise to magmatism may be understood, it is necessary to be able to "see through" the shallow-level processes which are responsible for much of the compositional diversity observed in volcanic arc magmas. It is relevant therefore to examine briefly the nature of differentiation processes and the relationships which evolved magmas have to their more primitive parents. To do this we have chosen a magma suite that undergoes extended differentiation from primitive basalt compositions through more evolved basalt, basaltic andesite, and andesite to dacite (Figure 3.8).

Compositional evolution of the magma suite in Figure 3.8 is governed largely by crystal fractionation. It involves the separation of evolving melt from a sequence of changing phenocryst assemblages, which are consistent in this case with the order in which they crystallize. The process of melt-solid separation is invariably driven by buoyancy difference between the melt and solid phases, such as in classical crystal settling/flotation models or the release of residual melt from cooler, crystallizing boundary layers at the margins of a magma body (McBirney et al., 1985; Marsh, 1989). Some magma compositions in Figure 3.8 are notable for cutting across bends in the compositional trajectories formed by fractional crystallization. These particular compositions reflect mixing between more- and less-evolved magmas, a process which may be confirmed by detailed petrographic examination and analysis of the phenocryst phases in such magmas. Indeed, it is common for magma mixing and crystal fractionation processes to combine, to varying extents, to produce the compositional variation observed in subduction zone magma suites (e.g., Sakuyama, 1981, 1984). Magma compositions may also be contaminated by xenogenous materials, either through complete or partial dissolution of solid wall rock or by admixture of already molten material. It is needless to say that there exists a diversity of potential contaminants within the crust and underlying lithosphere beneath volcanic arcs and that magma contamination is observed in some cases (e.g., Davidson, 1987; Gamble et al., 1993); however, consensus has it as an exception rather than the rule.

It is beyond our scope to cover the topic of differentiation in more detail. Suffice it to say that there exists an extensive literature dealing with both theoretical and observational aspects of the chemistry and physics of magma differentiation.

Primary Magma Definition

A primary magma is one unmodified in composition from that produced in the magma source region. Although this concept is simple, it is relatively difficult to identify precisely a primary magma composition. For example, in the case of mantle-derived magmas several factors contribute uncertainty, including plausible variations in the mantle source composition, the extent of equilibration achieved between the melt and residual mantle, and the mechanism of melt segregation, among others. A simple and widely used criteria for identifying potential mantle primary compositions is the FeO*/MgO weight ratio (FeO* is total iron as FeO) or Mg# ($100 \times \text{Mg}/(\text{Mg} + \text{Fe})$) of a magma. The composition of a melt formed in equilibrium with Mg-rich olivine, the main constituent of mantle peridotites, is constrained by the now well-established behavior of Fe-Mg partitioning between olivine and silicate melt at both atmospheric and higher pressures (Roeder and Emslie, 1970; Ford et al., 1983; Takahashi and Kushiro, 1983):

$$K_{d\,\text{olivine-liquid}}^{\text{Fe-Mg}} \equiv \left(\frac{\text{Fe}}{\text{Mg}}\right)_{\text{olivine}} \Big/ \left(\frac{\text{Fe}}{\text{Mg}}\right)_{\text{liquid}} = 0.3$$

Accepting this value and that the Mg# of mantle olivine is ~90, and if we assume a Fe^{3+}/Fe^{2+} ratio of 0.1 for the liquid phase, a mantle-

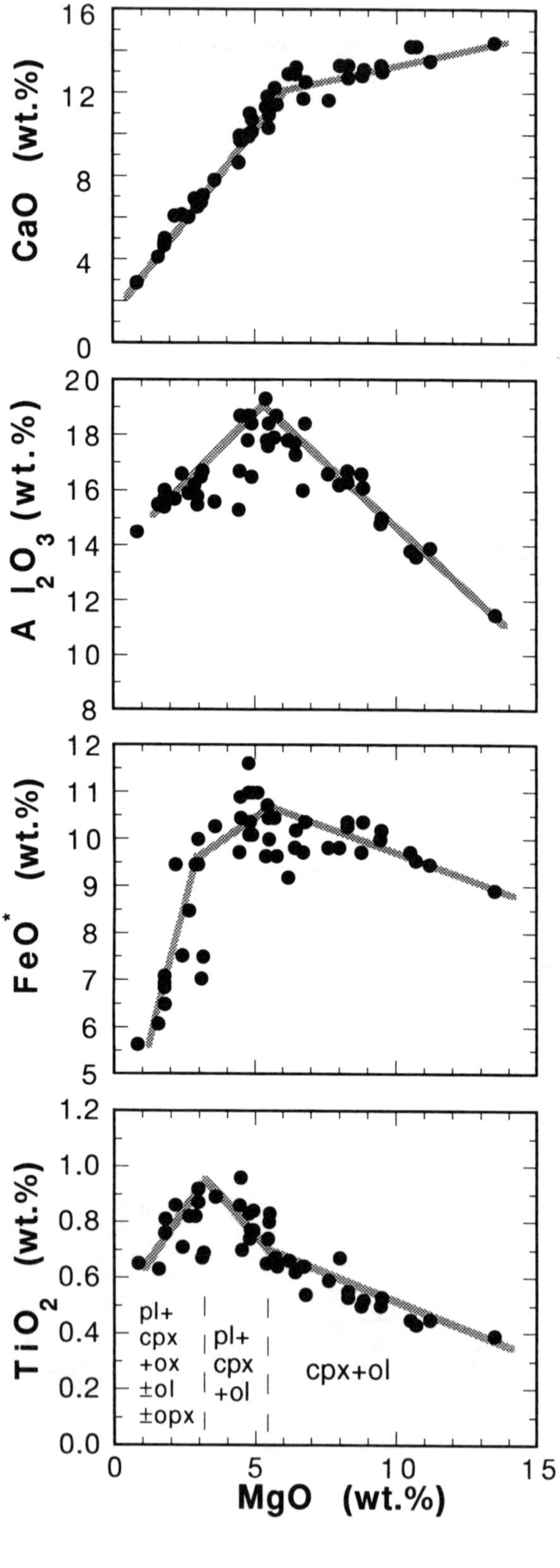

CaO (wt.%)
Al2O3 (wt.%)
FeO* (wt.%)
TiO2 (wt.%)
MgO (wt.%)
pl+
cpx
+ox
±ol
±opx
pl+
cpx
+ol
cpx+ol

derived primary magma is required to have a FeO*/MgO ratio less than unity or a Mg# greater than 70.

Both the FeO*/MgO ratio and Mg# change rapidly with any differentiation, since Mg-rich phases (olivine and pyroxenes) invariably dominate the early stages of magma crystallization. In the event of differentiated magma compositions not meeting the requirements for equilibrium with mantle peridotite, it may be possible to estimate primary magma compositions by accounting for the effects of prior fractional crystallization. This necessitates back addition of fractionated phases until a melt composition is capable of equilibrium with mantle olivine, as in the olivine maximum-fractionation model employed by Tatsumi and Ishizaka (1982a).

This method of estimating primary magma compositions necessarily assumes a known Mg# for mantle olivine. Although this is rather constant at around 90, it can vary by 2–3 units. To overcome this uncertainty, it may be possible in some cases to obtain further constraints from the compositions of the most magnesian phenocrysts observed in a magma suite (e.g., Nye and Reid, 1986) or to combine additional exchange partition coefficients for other elements between olivine and silicate liquid. Takahashi et al. (1987) have used Ni-Mg, Mn-Mg, and Fe-Mg exchange equilibria, combined with knowledge of the concentrations of these elements in mantle olivines to estimate the composition of primary magma. A similar but more qualitative guide is provided by the abundance of Ni and Cr in a magma, since these elements are rapidly depleted during the early stage of fractional crystallization because they are concentrated in olivine and chromite, respectively. Volcanic rocks with Ni content greater than 200 ppm and Cr content greater than 400 ppm are candidates for primary magmas (e.g., Sato, 1977; Hart and Davis, 1978; Kinzler et al., 1990). Finally, as a note of caution, the effects of crystal accumulation (or adding back too much olivine and chromite) need to be avoided when applying these methods to identify primary magma compositions.

Figure 3.8 Selected major element differentiation trends from primitive basalts to andesite and dacite, for Western Epi in the Vanuatu arc. Phases crystallizing in these magmas and determining the differentiation trend are plagioclase (pl), clinopyroxene (cpx), Fe-Ti oxide (ox), olivine (ol), and orthopyroxene (opx). *(Data reproduced by permission from Barsdell and Berry, 1990.)*

Subduction Zone Primary Magma Compositions

Magma suites which contain primitive compositions capable of equilibrium with mantle peridotite (i.e., with FeO*/MgO < 1) are comparatively uncommon in subduction zone settings. Nevertheless, examples do occur and have been described from virtually every subduction zone system—some notable occurrences, among others, being reported from the Solomon Islands (Ramsay et al., 1984), New Britain (Johnson et al., 1985), NE Japan (Yamamoto, 1988), the Aleutian Islands (Nye and Reid, 1986), and Vanuatu (Barsdell and Berry, 1990; Eggins, 1993). A more widespread occurrence for such primitive subduction zone magmas is supported by the presence of magnesian olivine and pyroxene (with Mg# between 88 and 94) and Cr-rich spinel as (1) phenocrysts in more evolved magmas (e.g., see Eggins, 1993; and locality compilation therein), (2) crystal cumulates brought to the surface by erupting magmas (e.g., Conrad and Kay, 1984; DeBari et al., 1987), and (3) cumulate sequences in the tectonically exhumed root zones of island arcs (e.g., DeBari and Coleman, 1989). Furthermore, primitive subduction zone magmas have been demonstrated to differentiate to the more evolved basalt (FeO*/MgO > 1), andesite, and dacite compositions that are more commonly observed in volcanic arcs elsewhere (e.g., Gust and Perfit, 1987).

High-Mg andesites (often referred to as boninites) are comparatively uncommon subduction zone magmas that also include primitive compositions capable of being in equilibrium with the mantle. They have been described from the forearc regions of the Japan, Mariana-Izu-Bonin, Aleutian, Vanuatu, and Tonga arc systems (e.g., Cawford et al., 1989; Falloon et al., 1989; Tatsumi and Ishizaka, 1982a,b; Murton et al., 1992). High-Mg andesites are also notable for their frequent presence in ophiolite sequences, with particularly good examples being described from Papua, New Guinea (Jenner, 1981), New Caledonia (Cameron, 1989), and SE Australia (Crawford et al., 1984). Primitive high-Mg andesites are characterized by extremely high levels of magnesian (Mg# > 90) phenocrysts of olivine and low-Ca pyroxene (sometimes clinoenstatite) or clinopyroxene and very-Cr-rich spinel.

We have compiled a brief list of previously published primary melt estimates and primitive lava compositions which span much of the range observed in subduction zones (Table 3.1). Cursory inspection reveals a diversity in composition that is unparalleled in its extent

Table 3.1 Possible primary magma compositions in subduction zones

	LOW K		MEDIUM K			HIGH K	SHOSHONITIC	HIGH-MG ANDESITE	
	RZ-6	IA-5	TM-0	ID1	AMBAE	OO-25	ABHK	TGI	CH414
SiO_2	49.09	49.68	50.57	48.66	48.44	49.21	49.80	59.59	57.60
TiO_2	0.65	0.53	0.51	0.49	0.58	1.03	1.71	0.44	0.26
Al_2O_3	15.09	15.45	14.48	13.45	11.79	15.18	17.54	13.55	11.06
FeO*	8.97	9.08	8.36	8.29	9.87	9.14	7.12	6.32	8.78
MnO	0.19	0.18	0.17	0.17	0.18	0.25	0.16	0.12	0.17
MgO	13.34	10.20	12.64	16.00	15.33	10.72	9.68	9.65	12.31
CaO	10.60	13.13	10.57	10.58	10.80	11.04	7.92	6.24	7.74
Na_2O	1.77	1.52	1.95	1.82	1.96	1.97	1.73	2.66	1.64
K_2O	0.20	0.17	0.65	0.47	0.89	1.41	3.56	1.30	0.42
P_2O_5	0.09	0.06	0.10	0.06	0.16	0.05	0.77	0.13	0.02
Total	100.00	100.00	100.00	100.00	100.00	100.00	100.00	100.00	100.00
FeO*/MgO	0.67	0.89	0.66	0.52	0.64	0.85	0.74	0.65	0.71

RZ-6 from Miocene NE Japan (Tatsumi et al., 1994b)
IA-5 from New Britain (Basaltic Volcanism Study Project, 1981)
TM-0 from Miocene NE Japan (Tatsumi et al., 1994b)
ID1 from Aleutian (Nye and Reid, 1986)
OO-25 from Quaternary NE Japan (Tatsumi, unpub.)
ABHK from SW Japan (Tatsumi and Koyaguchi, 1989)
TGI from Miocene SW Japan (Tatsumi and Ishizaka, 1981)
CH414 from Bonin Islands (Umino and Kushiro, 1989)

compared with that in many other tectonic settings. All the compositions have as a group low FeO*/MgO ratios suitable for equilibrium with the mantle, original lavas which contain highly magnesian phenocryst assemblages, and have in several cases been experimentally demonstrated to be consistent with melting of peridotite mineral assemblages (olivine + orthopyroxene ± clinopyroxene) at upper mantle pressures and temperatures. Based on these primary magma estimates, it can be concluded that the principal site of magma generation within subduction zones is in the mantle wedge.

A number of other primary magma types have been proposed or identified in subduction zones, of which the most significant are magma compositions consistent with melting of the subducting ocean crust, and for completeness we should also include melts formed in the volcanic arc crust itself. Possible primary melts of the subducted oceanic crust are similar to typical subduction zone dacites in composition, but are distinguished by relatively high Al_2O_3 contents (>15 wt% at 70 wt% SiO_2), high Ce/Yb, and low Y, Yb, and Sc concentrations (Drummond and Defant, 1990). They have been identified from subduction zone settings where very young and hot oceanic crust is subducted and are commonly referred to as adakites.

3.1.4 Trace element behavior

Trace elements are frequently used to quantify magmatic processes via either inverse or forward modeling strategies. They are also often used as qualitative discriminants between magma types, particularly between magmas erupted in different tectonic settings (for an overview, see Wilson, 1989, pp. 32–34).

The behavior of trace elements during partial melting processes is usually described by equilibrium batch melting or integrated fractional (Rayleigh) melting equations (Shaw, 1970). These equations, although based on static melting models, are robust and can be applied to modern concepts of decompression melting with dynamic melt segregation via pervasive porous flow or channeled melt flow (e.g., Iwamori, 1992).

They are formulated for either modal or nonmodal melting, depending on whether the solid phases melt in the proportions in which they are present in the source or in different proportions. Note that nonmodal melting is the usual case, particularly for mantle melting. The modal batch and fractional melting equations are

$$C_L = \frac{C_s}{D + F(1-D)}$$

and

$$C_L = \frac{C_s}{F}\left[1 - (1 - F)^{\frac{1}{D}}\right]$$

For nonmodal melting,

$$C_L = \frac{C_s}{D + F(1 - P)}$$

and

$$C_L = \frac{C_s}{F}\left[1 - \left(1 - \frac{PF}{D}\right)^{\frac{1}{P}}\right]$$

where C_L and C_S are the element concentrations in the partial melt and source material, F is the degree of melting or the mass fraction of the partial melt relative to the initial solid mass, D is the bulk partition coefficient given by

$$D \equiv \sum_{\alpha} X_{\alpha} D_{\alpha} \qquad \sum_{\alpha} X_{\alpha} = 1$$

where X_{α} are the phase fractions in the source and D_{α} are their respective solid-melt partition coefficients, and P is the equivalent coefficient calculated using the phase fractions which enter the melt (P_{α}):

$$P \equiv \sum_{\alpha} P_{\alpha} D_{\alpha} \qquad \sum_{\alpha} P_{\alpha} = 1$$

Trace element behavior in magmatic systems undergoing crystal fractionation can be described by the equilibrium and fractional (Rayleigh) crystallization equations, depending on whether the crystallizing solid phases remain in equilibrium with the evolving melt or become chemically and/or physically isolated from it. These equations are, respectively,

$$C_L = \frac{C_{L_0}}{D + F(1 - D)}$$

and

$$C_L = C_{L_0} F^{(D-1)}$$

where C_L and C_{L_0} are the element concentrations in the evolving and parent melts, D is a bulk partition coefficient, based on the crystallizing/fractionating phase assemblage, and F is the mass fraction of original melt remaining. Both equations produce similar results for incompatible elements except at very large extents of crystallization (i.e., $F \gg 0.5$).

A comparison of the trace element behavior produced by varying extents of partial melting and fractional crystallization for D values ranging from compatible ($D > 1$) to highly incompatible ($D \ll 1$) is shown in Figure 3.9. This illustrates a number of important concepts often applied by geochemists. The first stems from the limited relative variation that occurs between incompatible elements (D values $< \sim 0.3$) with up to moderate degrees of crystallization ($0.5 < F < 1$), particularly between elements which have similar D values. It follows that incompatible element ratios will still reflect primary melt values even given moderate extents of differentiation. The second relates to the unavoidable increase in incompatible element concentrations that occurs with differentiation. However, this variation is insignificant relative to that produced by partial melting, which is typically limited to F values less than 0.2–0.3 in the case of mantle melting. Even small variations in degree of melting will far outweigh variation arising from moderate degrees of crystallization. Lastly, note that magma mixing also has little effect on such element ratios unless it involves highly differentiated magmas.

In reality, D values may change during the course of differentiation, due to changes in the crystallizing phase assemblage (as illustrated in Figure 3.9), with elements becoming either more or less compatible. Examples of this behavior, which are important to the study of subduction zone magmas, occur with the onset of plagioclase and/or magnetite crystallization, as these phases preferentially incorporate key petrogenetic elements (i.e., Sr and Eu^{2+} in plagioclase and Ti and V in magnetite) which are otherwise incompatible during earlier olivine and/or pyroxene crystallization. It follows that concentrations and ratios of incompatible trace elements in subduction zone basalts which have not fractionated plagioclase or magnetite will approximate

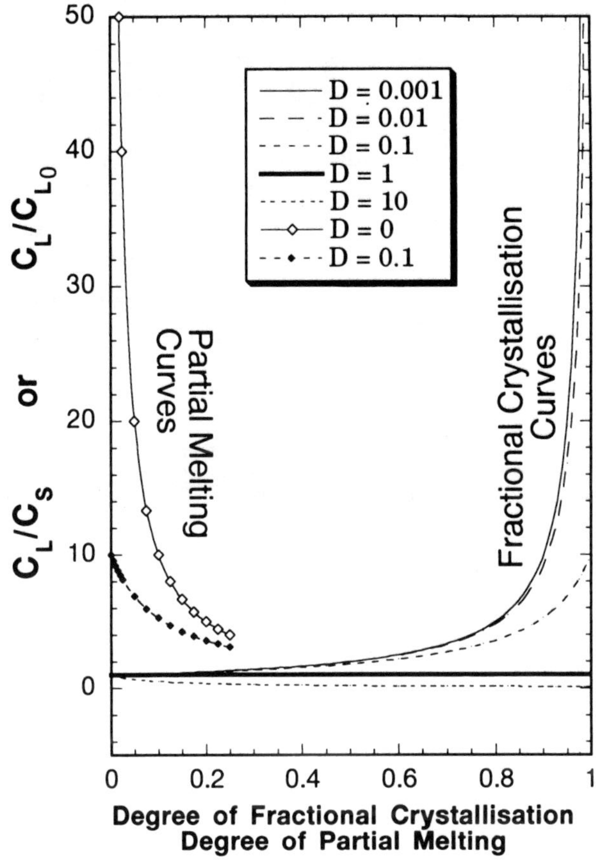

Figure 3.9 Concentration variation of trace elements with differing bulk partition coefficients (*D* values) during partial melting and fractional crystallization. Concentrations are shown relative to the source region for partial melting and relative to the initial melt for fractional crystallization. See text for details.

the concentrations and reflect the ratios of their parental primary magmas.

3.1.5 Trace element systematics in primitive subduction zone magmas

A compilation of trace element data has been made for subduction zone basalts with MgO > 5 wt% (equivalent to FeO*/MgO < ~2), a value chosen to avoid, as far as possible, the effects of plagioclase and magnetite fractionation. This compilation encompasses a considerable range in both incompatible element ratios and concentrations, spanning low-K to very-high-K compositions (Figure 3.10). Note that La is used in Figure 3.10 because it is the most commonly analyzed of the more incompatible trace elements for which good-quality data are available. Systematic variation of certain key trace element ratios occurs with changing incompatible element concentration, both globally and within individual subduction zones. This observation applies both to element ratios that can be expected to vary with differential degrees of melting, due to different element incompatibilities (e.g., La/Yb, Ti/Zr), and those which are typically invariant during mantle

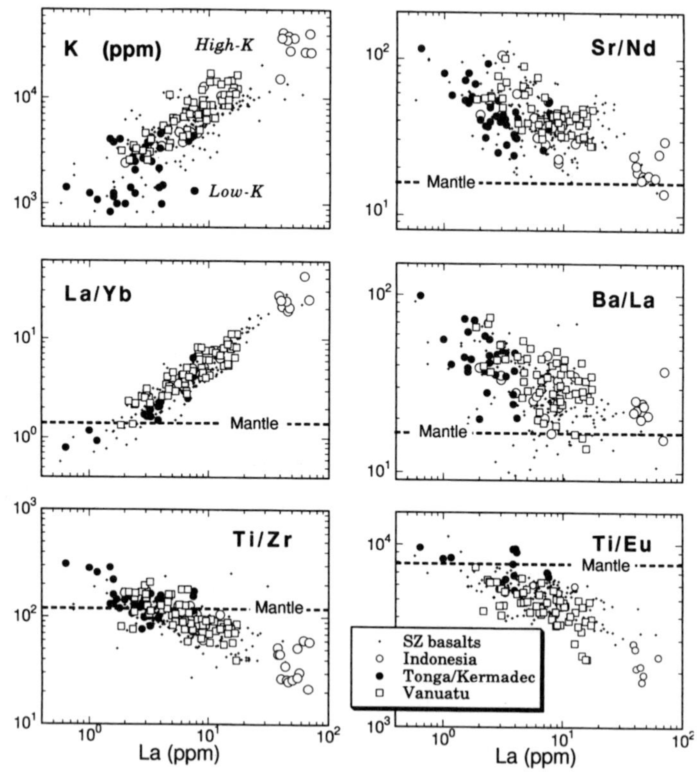

Figure 3.10 Examples of trace element systematics occurring in subduction zone basalts with MgO > 5 wt% or FeO*/MgO < 2. Mantle values for elemental ratios (as indicated by lines) are the primitive mantle values listed by Sun and McDonough (1989). Note that most OIB and MORB plot on or near the mantle value lines. *(Sun and McDonough, 1989.)*

melting, due to equivalent element incompatibilities. Several of the latter are plotted in Figure 3.10 (i.e., Sr/Nd, Ba/La, and Ti/Eu). The relative incompatabilities which apply to these and other elements during mantle melting can be gauged readily from Figure 3.10, where increasing incompatibility occurs toward the left (i.e., Yb to Cs) and elements with similar incompatibilites are placed side by side. A number of important aspects regarding trace element systematics in primary subduction zone magmas are to be gained from Figure 3.10, particularly the following:

1. Geochemical variability is more or less linear between two end-member compositions.
2. LILE/REE (large-ionic-radius lithophile element/rare-earth element) ratios are much higher than mantle values (and magmas typically derived therefrom, i.e., MORB and OIB) and decrease broadly with increasing La concentration.
3. HFSE/REE ratios tend to be lower than mantle values (and also mantle-derived magmas), and of these ratios Ti/Eu is notable for its systematic decrease below mantle values with increasing La concentration.

Basalt compositions representative of different regions within the compositional spectrum have been selected from various arcs and are plotted in multielement diagrams in Figure 3.11. These diagrams employ two element ordering schemes, one based on increasing ionic radius of elements and the other on relative element incompatibility during mantle melting (following the order established by Sun and McDonough, 1989). In both diagram types the element concentrations are divided by their respective concentrations in an average N-MORB (after Sun and McDonough, 1989), since this enables direct comparison with magma compositions formed by melting of the upper mantle in the absence of subduction. In addition to the systematics revealed in Figure 3.10, these diagrams show that many subduction zone magmas attain very low HFSE and REE concentrations relative to MORB, particularly the low-K magmas.

Origin of Trace Element Systematics

The diagnostic and elevated LILE/REE ratios of subduction zone magmas, which are particularly well developed in the incompatible element-poor, high-LILE/LREE, low-K lavas, are widely attributed to the

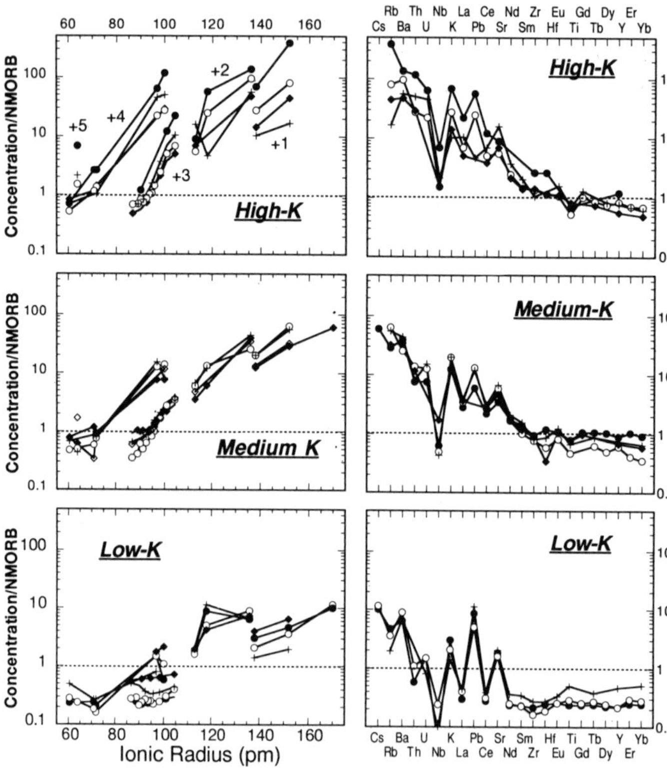

Figure 3.11 Multielement diagrams for primitive basalts selected from different magma series within the observed subduction zone compositional spectrum. Element concentrations are normalized to those in N-MORB (values from Sun and McDonough, 1989) and are plotted in order of increasing ionic radius (picometers) on the left-hand-side diagrams and order of decreasing incompatibility during mantle melting (as determined by Sun and McDonough, 1989) on the right-hand-side diagrams. Positive values denote ionic charge of the element groups. *[Data reproduced by permission from Bailey, et al. (1989), Ewart and Hawkesworth (1987), Woodhead and Johnson (1993), Gorton (1977), van Bergen et al. (1992) White and Dupre (1986).]*

addition of an aqueous fluid phase to their mantle wedge source. This assertion is supported by the existence of broad correlations between the size of an element's relative enrichment over other elements which have smaller radii and/or higher valence states and by experimental data which indicate increasing element mobility in aqueous fluids with decreasing ionic potential or charge-to-ionic-radius ratio

(Tatsumi et al., 1986; Chapter 6). Solid-melt partitioning is unlikely to account for these trace element characteristics, in particular the high LILE/LREE ratios, given existing knowledge of mineral-melt partition coefficients and the mineralogy of the subducting oceanic crust or mantle wedge. Finally, none of the components, or combinations thereof, which may be present within subduction zones match the observed magma trace element characteristics; hence, models invoking the bulk addition of such components, for example subducted sediment, to the magma source region are inconsistent with the data.

The very low HFSE and REE concentrations that characterize the low-K (low-La) magmas are unable to be produced by plausible degrees of melting (0–20% or even 30%) of a normal upper mantle (MORB-type) source. These low concentrations are most readily attributed to their mantle wedge source having been depleted in incompatible elements by prior melting events, as argued by Woodhead et al. (1993), who demonstrated this to be consistent with HFSE systematics in primitive arc magmas, such as the occurrence of Ti/Zr ratios greater than MORB and the MORB mantle source. These authors further suggested, along with Ewart and Hawkesworth (1987) and others, that the more geochemically depleted the mantle wedge source the more susceptible it may be to displaying the effects of metasomatic enrichments derived from the subducting oceanic lithosphere.

High-K (high-La) subduction zone basalts are notable for having incompatible trace element signatures and elemental ratios distinct from those of the low-K (low-La) basalts, particularly the development of a strong Ti depletion relative to REE. This particular depletion has been a focus of conjecture as to the possible presence of a residual, Ti-rich, accessory phase (e.g., rutile) during the subduction zone melting process. Green and Pearson (1986) and Ryerson and Watson (1987) have shown that basaltic mantle melts such as the high-K primary magmas are unlikely to be saturated with a residual titanate phase at their source, but that very silica-rich melts of the oceanic crust may be saturated with such phases. More recently, Ayers and Watson (1993) have demonstrated that the solubility of TiO_2 in aqueous fluids decreases markedly with increasing pressure and that high-pressure aqueous fluids have a limited capacity for transporting TiO_2. Given the distinct LILE/LREE ratio characteristics, LREE-rich nature, and relatively large TiO_2 depletion observed in high-K subduction zone basalts, the chemistry of these magmas may reflect addition of a second, higher-pressure (?) aqueous fluid phase, or, perhaps, even a silicate melt derived from the subducting ocean crust, to their mantle wedge source region.

The systematic geochemical variation occurring within primitive subduction zone basalts is broadly consistent with the following simple model. Fluids derived from dehydration of the subducting oceanic lithosphere preferentially transport elements with small ionic potential from the subducting ocean crust to the mantle wedge and ultimately the magma source region. The geochemical characteristics of these fluids are reflected variably in subduction zone magmas in accordance with relative fluid/melt ratios and the trace element component in the melt derived from the unmetasomatised mantle wedge. The latter may vary significantly as a function of degree of melting and mantle wedge composition. The fluid signatures will be best developed where mantle wedge material is more depleted in incompatible elements, and least where small amounts of fluid combine with small degrees of melting of undepleted mantle wedge material. Alternatively, the two compositional end members of subduction zone basalts could reflect the addition of two geochemically distinct fluids, or a fluid and a SiO_2-rich silicate melt, to the mantle wedge source region.

3.1.6 Radiogenic isotope tracers

The radiogenic isotopes of Sr, Nd, Pb, and, to a lesser extent, Hf are sensitive tracers of mantle wedge and subducting oceanic lithosphere contributions to subduction zone magmas. This stems from the long half-lives of their decay schemes compared with the time scales of subduction processes and from the large contrast in isotopic composition and element concentrations that often exist between the various possible contributing components to subduction zone magmas (e.g., depleted upper mantle, subducted altered oceanic crust, and sediment). The effective use of these radiogenic isotopes requires that the isotopic composition of all possible contributing source components be known and that they display sufficient compositional contrast. Difficultics invariably arise on both accounts, due to uncertainties in constraining the isotopic composition of the mantle wedge (unmodified by subduction) and to the sometimes similar isotopic composition of subducted sediments, the volcanic arc crust, and particular enriched mantle materials that comprise the source of some hot-spot magmas and which may be present in the mantle wedge.

White and Patchett (1984) have shown that the isotopic compositions of Sr, Nd, Pb, and Hf in subduction zone magmas overlap to only a limited extent with MORB, with the exception of Hf. This indicates that the mantle wedge magma source does not comprise MORB-type mantle, as is believed to typify the upper mantle, or that MORB-type mantle material in the mantle wedge has been largely overprinted by subduction effects (barring magma contamination during passage through the subarc crust or lithosphere).

$^{87}Sr/^{86}Sr$ versus $^{143}Nd/^{144}Nd$ variation in magmas from various subduction zones can be divided into two broadly different types of behavior. A-type rocks occurring in oceanic subduction zones tend to display horizontal or even positive trends and are restricted to relatively low $^{87}Sr/^{86}Sr$ and high $^{143}Nd/^{144}Nd$ values. B-type rocks from continental subduction zones display negative trends that extend from the field of oceanic arcs to very high $^{87}Sr/^{86}Sr$ and low $^{143}Nd/^{144}Nd$ values (Figure 3.12). This division corresponds with that made by Hawkesworth et al. (1993), who distinguished low Ce/Yb (oceanic) and high Ce/Yb (continental) subduction zones. Shown for comparison in Figure 3.12 are the approximate compositional fields of MORB (representative of the upper mantle), and part of the field for oceanic sediments. Not shown is the compositional field of OIB, which corresponds fairly closely with the array of continental arc compositions, beginning near the MORB field and fanning out beyond the bulk-earth composition, and altered ocean crust (which has $^{87}Sr/^{86}Sr$ elevated over MORB values through addition of radiogenic seawater Sr). In this diagram, those subduction zone magmas which have high elemental Sr/Nd values occur either in the oceanic subduction zones or at the "depleted" (low $^{87}Sr/^{86}Sr$) ends of the continental subduction zone arrays. It may also be significant that these particular magmas tend to plot on the high $^{87}Sr/^{86}Sr$ side of the mantle array toward the composition of altered oceanic crust, suggesting the elemental Sr excess in such magmas is sourced via aqueous fluids from seawater-altered oceanic crust. That some of these magmas have $^{143}Nd/^{144}Nd$ lower than MORB and altered ocean crust estimates requires an additional unradiogenic Nd component that is probably derived from either sediment or an isotopically enriched OIB mantle component present in the

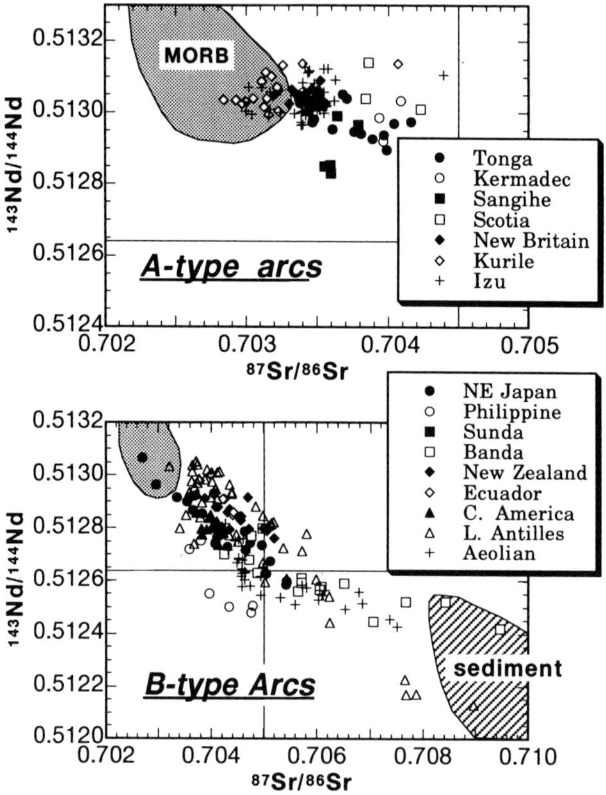

Figure 3.12 Sr-Nd-isotope characteristics in subduction zone basalts. *(Data reproduced by permission: for Tonga, Kermadec, and New Zealand: Ewart and Hawkesworth, 1987; for Sangihe: Tatsumi et al., 1991; for Scotia: Cohen and O'Nions, 1982; for New Britain: Woodhead and Johnson, 1993; for Kurile: Zhuravlev et al., 1987; for Izu: Tatsumi et al., 1992; for NE Japan: Nohda and Wasserburg, 1981, 1986; for Philippines: Defant et al., 1988; Defant et al., 1989; for Sunda and Banda: Wheller et al., 1987; Stolz et al., 1988; Edwards et al., 1991; for Lesser Antilles: Davidson, 1986; for Aeolian: Ellam et al., 1989.)*

mantle wedge. The positive $^{87}Sr/^{86}Sr$ versus $^{143}Nd/^{144}Nd$ arrays observed by Tatsumi et al. (1991) for the Sangihe arc have been cited as evidence for an enriched OIB mantle source in that particular subduction zone. The isotopic composition and systematics of the continental subduction zones are consistent with either a large and variable sediment contribution to their source regions or contamination by high $^{87}Sr/^{86}Sr$ and low $^{143}Nd/^{144}Nd$ materials possibly during ascent through the continental crust.

Individual subduction zone systems typically exhibit steep arrays on $^{208}Pb/^{204}Pb$ and $^{207}Pb/^{204}Pb$ versus $^{206}Pb/^{204}Pb$ plots, intermediate between MORB and, or overlapping with, local sediments on the subducting plate (e.g., Woodhead and Fraser, 1985; Hawkesworth et al., 1993). This has yielded strong support for incorporation of subducted sediment-derived Pb into the subduction zone magma source. However, this conclusion is ambiguous where the volcanic arc crust may contain a "radiogenic Pb component" as in the case of continental arcs. Such ambiguities have proved difficult to resolve, as evidenced by conflicting models for the origin for isotope systematics in parts of the Lesser Antilles arc (White and Dupre, 1986; Davidson, 1987). As an added note on this point, we point out the dramatic change from oceanic to continental arc isotope characteristics with progression across the oceanic-continent boundary along the continuous Kermadec–Taupo volcanic zone arc system (Gamble et al., 1993).

3.1.7 Other elements and isotopes

The light elements (Li, Be, and B) and their isotopes are being used increasingly as tracers of sediment and altered ocean crust contributions to subduction zone magmas and as indicators of fluid involvement in subduction zone magmas genesis. The cosmogenic isotope ^{10}Be has provided the only undisputed evidence for a subducted sediment contribution to subduction zone magmas (e.g., Morris and Tera, 1989). ^{10}Be, which is formed in the atmosphere when cosmic rays break apart atoms of oxygen and nitrogen, accumulates in the upper layers of oceanic sediments. ^{10}Be is radioactive with a half-life of 1.5 million years and can be used to trace subducted sediments through subduction zone systems, provided the time to complete the cycle from formation to eruption is less than around 10 million years.

B is a good elemental tracer of slab-derived components because it is highly concentrated in sediments and altered ocean crust basalts (~100 ppm in both) relative to the mantle (~0.1 ppm; Ryan and Langmuir, 1993). In addition, B is known to be highly mobile in fluids at elevated pressure and temperatures, as evidenced by B depletion in progressively metamorphosed rocks (e.g., Bebout et al., 1993), and the subducted-sediment-derived flux of B that has been observed to an accretionary prism in a forearc setting (You et al., 1993). It is not surprising therefore that B concentrations are enriched in subduction zone magmas relative to MORB and OIB and relative to elements which behave similarly during partial melting processes (i.e., the LREE and Be). Ryan and Langmuir (1993) concluded that the high B/Be ratio values (~4 to ~100) of subduction zone magmas compared to MORB and OIB (<4) are a consequence of preferential transport of B in aqueous fluids derived from the subducting ocean crust. Ishikawa and Nakamura (1994) have since further demonstrated, using B isotopes, that the B present in subduction zones magmas from the Izu arc is derived specifically from altered oceanic crust rather than subducted sediments.

Several short-lived isotopic decay products of ^{238}U have provided (and continue to provide) insights into the origin and timing of certain element fractionations observed in subduction zone magmas (e.g., Gill and Williams, 1990; McDermott and Hawkesworth, 1991). Considerable data has now been accumulated for ^{230}Th, a decay product of ^{238}U which has a half-life of ~75,000 years, and a more limited amount of data for ^{226}Ra (a chemical analogue of Ba) which has a half-life of ~1500 years. Events which fractionate the elemental ratios Th/U and Ra/Th can be traced on time scales up to approximately five times these half-lives. These data reveal that some subduction zone magmas have undergone fractionation of U relative to Th and Ra relative to Th, well within the last 300,000 and 8000 years respectively. The nature of the processes which may cause these fractionations remains equivocal (see Gill and Williams, 1990); however, in the simplest model, they are interpreted to reflect relatively rapid melting and melt segregation following introduction of a Ra- and U-enriched fluid to the magma source region.

3.1.8 H_2O content of subduction zone magmas

A great deal of direct and circumstantial evidence has been cited for elevated H_2O contents in subduction zone magmas, including the

- Crystallization of hydrous phases (e.g., hornblende)
- Explosive nature of many eruptions and the obvious volatile outgassing that takes place from many subduction zone volcanoes
- Occurrence of mantle melting in a region cooled by subduction of the cold oceanic lithosphere and where the mantle solidus temperatures may not otherwise be exceeded

In contrast, few measurements of H_2O concentrations (and other volatile components) in subduction zone magmas have been made.

Direct measurements have been obtained by infrared spectroscopy on deeply erupted submarine glasses from the Vanuatu and Tongan arcs by Danyushevsky et al. (1993) and by ion-microprobe analysis of glass inclusions enclosed in phenocrysts from various localities (Sisson and Layne, 1993). H_2O contents in low- to medium-K basalt and basaltic andesite glasses from the Vanuatu and Tongan ranges between 1.2 and 3 wt%, though comparatively shallow eruption depths indicate likely vapor saturation and possible H_2O loss in some cases. These glasses are likely to be the products of approximately 50% crystallization; hence, because H_2O is essentially incompatible during crystallization, primary H_2O concentrations can be predicted to range from 0.6 to 1.5 wt%, though possibly exceeding 1.5 wt% if degassing has taken place. Measurements on high-Mg andesites dredged from the same region range between 1.3 and 2.5 wt% H_2O, but many of these values lie near or above vapor-saturated conditions at their depths of eruption. Measurements on basaltic glass inclusions in olivine phenocrysts from Fuego volcano, Guatemala, range between ~4 and 5.5 wt% and require relatively high primary values, around 2.5 to 4 wt%, assuming the melt inclusions are the products of 30–40% crystallization. In contrast, measurements on similar olivine-hosted glass inclusions from the Cascades reach only 3 wt% H_2O and indicate primary contents in the range 0 to 2 wt%.

Some studies have estimated H_2O contents using less direct methods. Sakuyama (1979) proposed that primary magmas in the NE Japan arc have between 0.5 and 3 wt% dissolved H_2O, based on the decrease in quartz liquidus temperature relative to that of hornblende and biotite, and the observed crystallization sequences of these phases in various magma suites. Knowledge of the differentiation of these magma suites enabled Sakuyama to place constraints on the values in their estimated primary melt compositions. Gaetani et al. (1993) have more recently proposed that the well-known liquidus depression of plagioclase crystallization is consistent with up to 6 wt% H_2O in certain arc lavas of basaltic andesite and relatively evolved basalt compositions. If their estimates are correct they imply concentrations of up to ~3–4 wt% H_2O in corresponding primary magmas.

It is clear that the H_2O contents of primary subduction zone magmas vary appreciably, given existing estimates range widely from 0 to ~4 wt%, but are in general considerably higher than the H_2O content which characterizes magmas in other tectonic settings, in particular N-MORB (~0.1–0.2 wt% H_2O). It is notable that most direct measurements indicate relatively low values between 0.5 and 2 wt% H_2O for low- to medium-K primary magmas in the Vanuatu arc, and that these values are considerably lower than those (i.e., 3–4 wt% H_2O) derived from estimates for similar magma compositions by Gaetani et al. (1993). This is a fertile area for further research which has significant implications for subduction zone magma genesis.

3.2 Across-Arc Variation in Magma Chemistry

In Chapter 2 we showed that in volcanic arcs where the angle of subduction is not steeply inclined the volcanic arc increases in width and two separate volcanic chains may be resolved. A remarkable characteristic of these arcs is the systematic variation in lava compositions that often occurs in the across-arc direction, in particular, an increase in K_2O content of magmas toward the backarc as depth to the subducting lithosphere increases. In this section we review evidence for the occurrence of across-arc variation in magma chemistry and phenocryst assemblages and investigate its origins.

Exceptions exist, as might be expected, to the "normal" across-arc geochemical variation that we shall exemplify using the NE Japan arc. Reverse patterns of K_2O variation have been demonstrated for lavas in the Banks Islands across the Vanuatu arc (Mallick, 1973; Ash et al., 1980; Barsdell et al., 1982) and in the Miocene NE Japan arc (Togashi, 1978; Tatsumi et al., 1988) (Figure 3.13). These and other exceptions have been a source of criticism as to the significance of the "normal" K_2O across-arc relationship (Johnson, 1976; Arculus and Johnson, 1978; Carr et al., 1979); however, we believe that valid explanations can be advanced for each of these cases, usually based on local tectonic complexities such as proximity to backarc rifting (as we indicate in Chapter 5).

3.2.1 Incompatible element concentrations

It has been repeatedly documented that concentrations of K and the LIL-group elements (i.e., Cs, Rb, K, Ba, and Sr) tend to increase with distance from the volcanic front, since first suggested by Tomita (1935). This constitutes the *K-h* relationship, whereby K_2O contents (K) of lavas at a fixed SiO_2 content correlate with depth (h) to the dipping seismic zone (Dickinson, 1975). An updated compilation of K_{51}-values (K_2O wt% at 51 wt% SiO_2 calculated from linear regression of SiO_2 and K_2O) confirms this relationship in Figure 3.13. In the NE

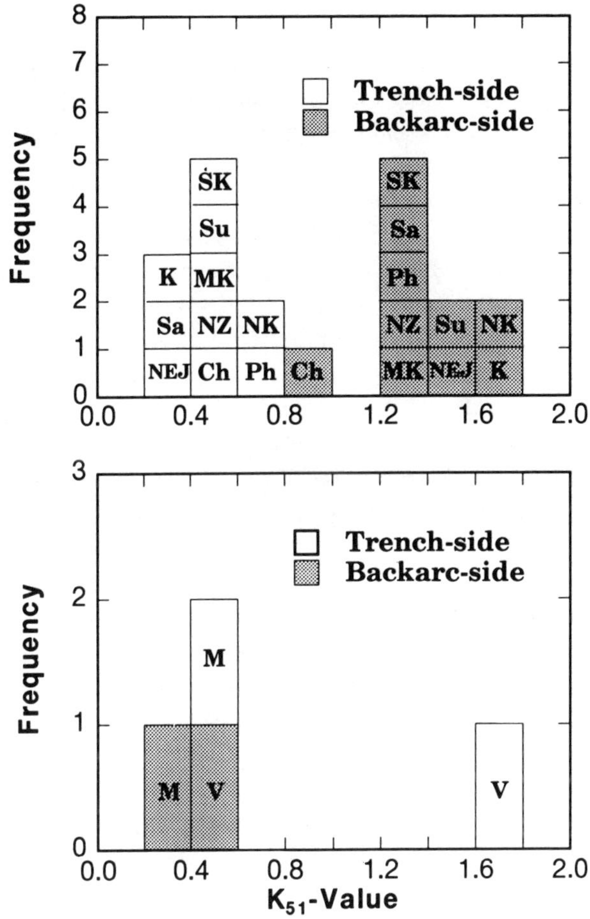

Figure 3.13 K_{51}-values for trench- and backarc-side volcanic chains in the following subduction zones: K = Kamchatka (Tatsumi, unpub. data), Sa = Sangihe (Tatsumi et al., 1991), NEJ = NE Japan (Sakuyama and Nesbitt, 1986), SK = South Kurile, MK = middle Kurile, NK = North Kurile (Bailey et al., 1989), NZ = New Zealand (Tatsumi and Tsunakawa, 1992), Ch = Chile (Hickey et al., 1986), Ph = Philippines (Defant et al., 1989), M = Miocene NE Japan (Togashi, 1978; Shuto, 1988), V = Vanuatu (Barsdell et al., 1982).

Japan arc, for which there exists a great deal of data on lava chemistry, across-arc variations have also been documented in other LIL elements, LREE, Hf, Zr, Nb, Th, U, and Pb (Sakuyama and Nesbitt, 1986). These particular elements are all incompatible with mineral-silicate melt distribution coefficients (*D* values) much smaller than unity.

Across-arc variation in incompatible element concentrations has also been associated with across-arc change in magma bulk composition, from tholeiitic through "calc-alkaline" to alkaline compositions with increasing distance from the volcanic front (Jakes and White, 1969, 1972; Jakes and Gill, 1970). To avoid the problematic nomenclature used by these authors, it can be demonstrated that across-arc variation also occurs in magma normative compositions. Basalts in the NE Japan arc, which have undergone similar degrees of differentiation and have FeO/MgO ratios of 1.5 ± 0.1, possess a higher normative *quartz* component in the trench-side volcanic chain than in the backarc-side chain, as illustrated by projections onto the normative planes, *diopside-olivine-quartz* from *plagioclase* and *diopside-olivine-quartz* from *diopside* (Figure 3.14). The proposed cause of these across-arc differences in incompatible element concentrations and normative composition encompass a diversity of mechanisms which shall be reviewed briefly below.

Crustal Thickness

A positive correlation between K_2O content, at a given SiO_2 content, and the thickness of crust was demonstrated by Condie and Potts

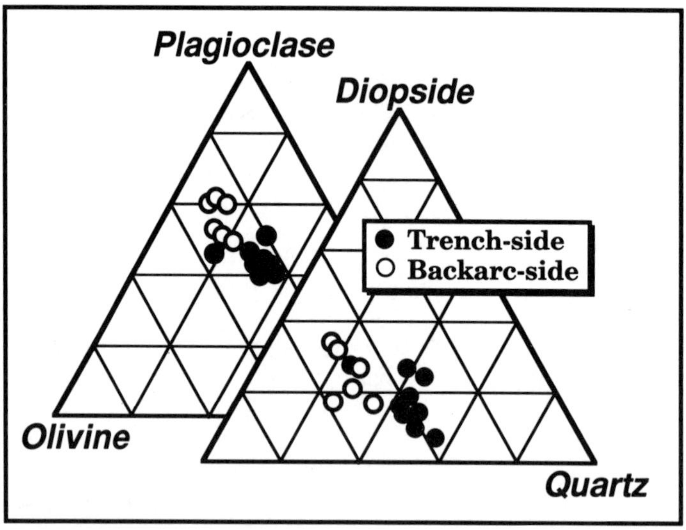

Figure 3.14 Normative composition projections (following Walker et al., 1979) of basalts with FeO*/MgO ratios of 1.5 ± 0.1 in the NE Japan arc. *(Data reproduced by permission from Kawano et al., 1961 and Sakuyama and Nesbitt, 1986.)*

(1969), based on a compilation of lava chemistry for 26 volcanic suites in the circum-Pacific region. Miyashiro (1974) has since suggested that the proportion of "calc-alkaline" (meaning higher K) rocks within a volcanic arc increases with thickening and development of continental-type crust. A negative correlation between Bouguer anomaly and K_{55}-value in the Japanese Islands, noted by Ui and Aramaki (1978), is consistent with this notion. These studies emphasized the role of crustal thickness largely in terms of explaining interarc difference in K_2O contents; however, the idea has been applied more recently to explain across-arc chemical variation (e.g., Meen, 1987). Thicker crust may contribute to increasing K_2O content in a magma in two ways: (1) by larger degrees of crustal contamination during rise through thicker crust (Condie and Potts, 1969, Leeman, 1983; Meen, 1987) and (2) by differences in liquid lines of descent resulting in more K-rich magmas through higher pressure crystallization at or near the base of the crust (Ui and Aramaki, 1978; Meen, 1987, 1990).

A role for crustal contamination in producing across-arc variation of incompatible elements can be challenged based on the following observations: (1) the thickness of the crust is constant (~33 km; Yoshii and Asano, 1972; Zhao et al., 1990) beneath the NE Japan arc; (2) across-arc variation is observed both in continental and oceanic arcs, where the compositions of crust may be quite different; (3) concentrations of Sr do not in general correlate positively with $^{87}Sr/^{86}Sr$ ratio in mafic lavas (e.g., Gill, 1981); (4) higher Sr concentrations and lower $^{87}Sr/^{86}Sr$ ratios characterize backarc-side lavas in NE Japan (Nohda and Wasserburg, 1981); and (5) primitive trench-side magmas are denser than backarc-side magmas in NE Japan, hence are less likely to pond in or at the base of the crust and facilitate crustal contamination (Tatsumi et al., 1983).

Ponding of magma at depth within volcanic arcs (e.g., at the Moho), prior to evolution of lower-density liquids that can rise more easily to the surface, may also control liquid lines of descent and incompatible element concentrations as a consequence of pressure effects on the crystallizing phase assemblage. Meen (1987, 1990) showed experimentally that the stability field of olivine shrinks against pyroxenes and plagioclase with increasing pressures and that a greater increase of incompatible elements is observed for higher-pressure magmatic differentiation. This mechanism may explain some interarc differences in incompatible element concentrations, but is unlikely to apply generally to across-arc variation, where there is (1) constant thickness of the crust beneath the arc, and in cases (2) where Mg-rich, least-differentiated lavas can be shown to have crystallized only olivine, as in the NE Japan arc.

Varying Source Composition

Across-arc variation in mantle wedge composition may account for variation in incompatible elements, with more K-rich sources occurring beneath more K-rich backarc-side volcanoes. However, this is a rather ad hoc remedy without a satisfactory explanation for why the source composition varies. Jakes and White (1970) have proposed that source regions bearing amphibole and K-rich phlogopite contribute to trench- and backarc-side volcanic chains, respectively. However, if source composition alone is the cause of across-arc variation in magma K contents, then elements which are more enriched in amphibole than phlogopite, such as sodium and samarium, might be expected to be higher in trench-side lavas, but the reverse is observed (e.g., Gill, 1981).

Difference in Degree of Melting

Variation in incompatible element concentrations are readily produced by varying the degree of partial melting of a source material, such that decreasing partial melting may explain across-arc variation trends. This has been tested by Sakuyama and Nesbitt (1986) for Quaternary volcanic rocks in the NE Japan arc, using a method proposed by Minster and Allegre (1978) and Hofmann and Feigenson (1983), based on the behavior of incompatible trace elements during batch partial melting. The relevant equation can be combined for two elements, x and y, to eliminate F and to obtain

$$\frac{1}{C_L^y} = \frac{C_S^x\left(1-P^y\right)}{C_S^y\left(1-P^x\right)}\frac{1}{C_L^x} + \left[\frac{D^y}{C_S^y} - \frac{D^x\left(1-P^y\right)}{C_S^y\left(1-P^x\right)}\right]$$

If a population of primary magmas is produced by different degrees of partial melting of a uniform source composition (i.e., the D and C_S values are identical for each magma), and quasi-eutectic melting applies (i.e., P is similar for all magmas formed), then the data set will form straight lines described by the preceding equation.

Concentrations of incompatible elements in primary magmas for NE Japan were back-calculated by Sakuyama and Nesbitt (1986) and their compositions plotted, using 1/Zr as the independent variable (see Figure 3.15). All incompatible elements, except Rb, were found to form straight lines, indicating that varying degrees of melting of a single source composition is consistent with the across-arc geochemical variation observed. The small deviation observed for Rb in trench-side magmas was attributed to either Rb heterogeneity in the source or a difference in the composition of the fluid phase being added to the mantle wedge across the volcanic arc. By assuming the initial concentration of an element in the source and bulk partition coefficients between the source and melt (for example, C_S^{Sr} = 50 ppm and D^{Th} = 0.001), Sukuyama and Nesbitt (1986) were able to calculate the degree

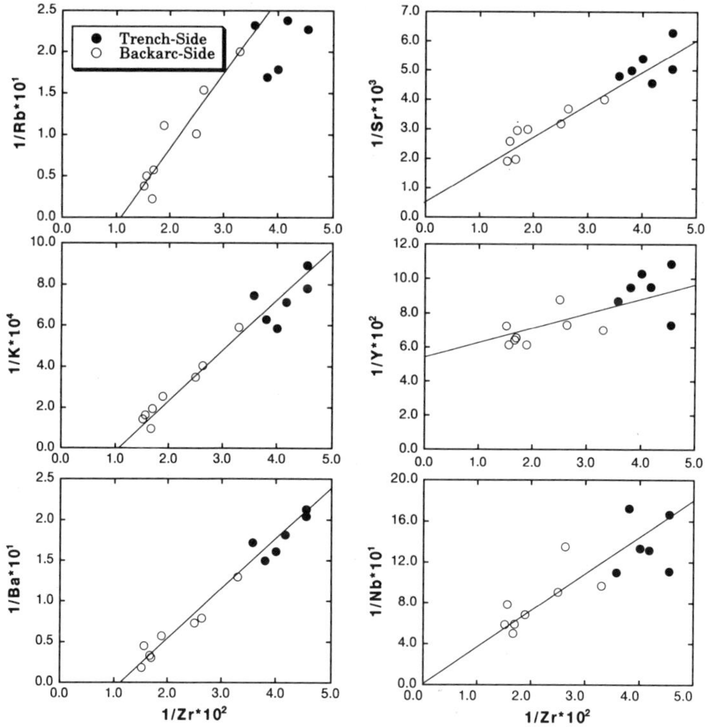

Figure 3.15 Incompatible trace element systematics for primary magma estimates in the NE Japan arc. *(Data reproduced by permission from Sakuyama and Nesbitt, 1986.)*

of partial melting for trench- and backarc-side magmas at 22.9% and 1.3% respectively.

Support for differing degrees of partial melting as the underlying cause of across-arc variation comes from a number of additional geochemical and experimental observations. The more LREE-rich REE patterns and greater relative HREE depletion in backarc-side magmas in numerous arcs (Figure 3.16) are consistent with the presence during melting of significant proportions of a phase or phases which have relatively high D_{HREE}, such as clinopyroxene ($D_{HREE} \sim 1$) and garnet

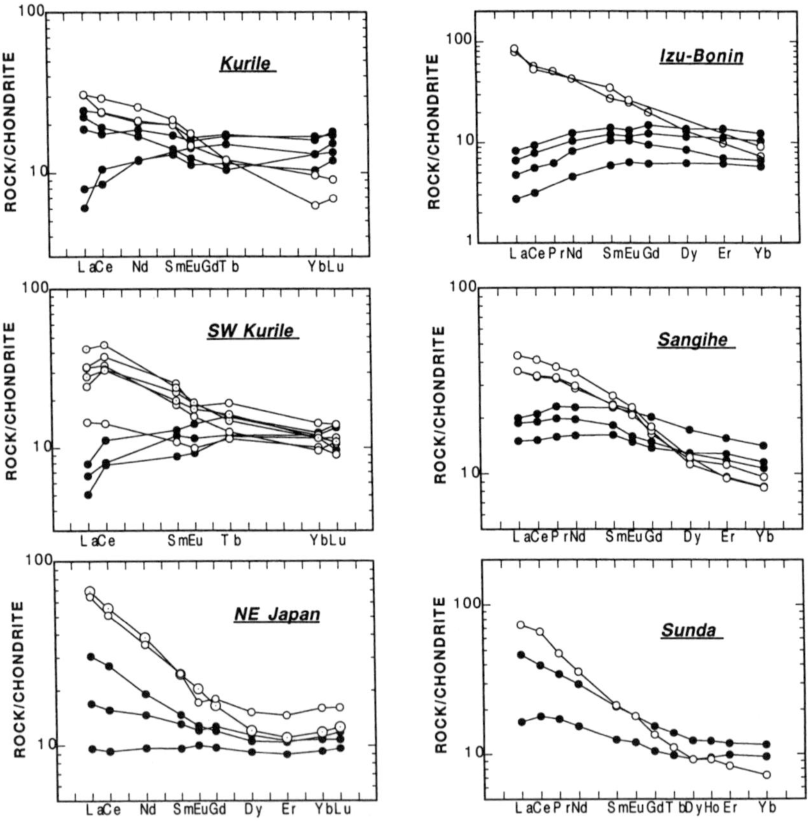

Figure 3.16 Chondrite-normalized REE patterns for basalts in the Sangihe (Tatsumi et al., 1991), Sunda (Hutchinson, 1982), Izu-Bonin (Tatsumi et al., 1992), NE Japan (Fujitani and Masuda, 1981), SW. Kurile (Masuda et al., 1975), and Kurile arcs (Bailey et al., 1989). Closed and open circles indicate trench- and backarc-side basalts, respectively.

($D_{HREE} \gg 1$). The presence of significant amounts of residual clinopyroxene and possibly also garnet during the formation of backarc-side magmas further suggests relatively small degrees of melting since these phases are preferentially removed from the source during melting (e.g., Mysen and Kushiro, 1977). High-pressure melting-phase relations of NE Japan primary melt compositions also suggest smaller degrees of partial melting toward the backarc (Chapter 5) as trench-side compositions are multiply saturated with olivine and orthopyroxene, whereas backarc-side compositions have clinopyroxene as an additional phase (Tatsumi et al., 1983).

What mechanism(s) could be responsible for decreasing degrees of melting from the volcanic front toward the backarc? Gill (1981) proposed that greater amounts of H_2O-rich fluid added to the magma source beneath trench-side volcanoes may promote greater degrees of melting there. This seems intuitively reasonable; however, degree of melting is a relatively complex function of many variables which, in addition to H_2O, include the pressure (P) and temperature (T) of melting. The dT/dP slope for the peridotite solidus is greater than that for adiabatic decompression of partially melting peridotite, and the contours for degree of partial melting lie almost parallel to the solidus at lower degrees of melting (Figure 3.17). It follows that higher degrees of melting can result from melting along a higher-temperature adiabat or by decompressing the mantle to shallower levels along a particular adiabatic ascent path. The interplay of H_2O, P, and T in determining the extent of melting and composition of subduction zone magmas will be investigated in more detail in Chapter 5.

3.2.2 H_2O content

Across-arc variation in phenocryst assemblages has been examined by Sakuyama (1977, 1979, 1983a), specifically to assess variation of H_2O contents in subduction zone magmas. Two types of different phenocryst assemblages occur in lavas and pyroclastic rocks in the NE Japan arc; magmas containing quartz phenocrysts but no hornblende are distributed along the trench side of the arc, and rocks containing hornblende phenocrysts but no quartz occur along the backarc side of the arc (Figure 3.18). Experimental data (Robertson and Wyllie, 1971; Maale and Wyllie, 1975; Wyllie, 1977) indicate (1) a large decrease in the liquidus temperature of quartz with increasing H_2O content in silicate melts and (2) a constant liquidus temperature for hornblende

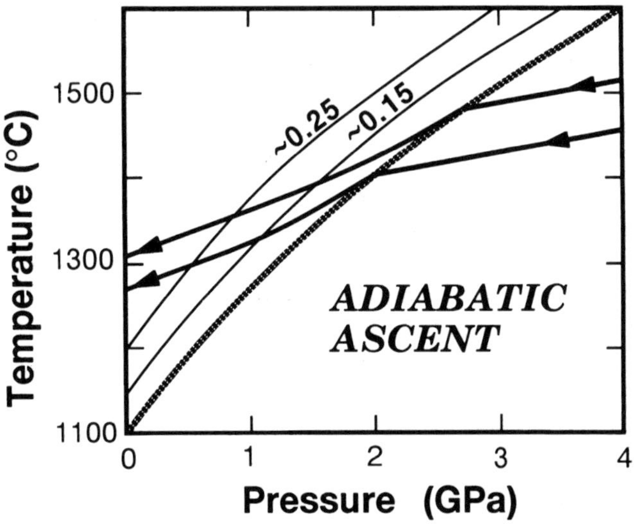

Figure 3.17 Schematic adiabatic ascent paths for mantle material passing through the peridotite solidus (hatched line). Thin lines indicate curves for equivalent (~25% and ~15%) degrees of partial melting.

(~950°C) in the presence of more than ~1 wt% H_2O. This enables a schematic isobaric T-X_{H_2O} section for andesite and dacite compositions to be drawn (Figure 3.19), which indicates that a magma crystallizing quartz prior to hornblende must contain a smaller amount of H_2O than if hornblende precedes quartz. Based on these experimental constraints and petrographic observations, Sakuyama concluded that H_2O contents are greater in the backarc-side magmas, and was able to show that an equivalent across-arc variation in phenocryst assemblages and H_2O content also occurred in the North Island of New Zealand. Estimated H_2O contents in NE Japan primary magmas (Aoki et al., 1981; Sakuyama, 1983a) indicate greater amounts of H_2O in backarc-side magmas (1–2 wt%) compared with trench-side magmas (<0.5 wt%).

Across-arc variation of H_2O in primary magmas has also been suggested based on F geochemistry (Ishikawa et al., 1980; Aoki et al., 1981). F partially replaces (OH) in hydrous minerals in the upper mantle, such as amphibole and phlogopite, because the F anion has an ionic radius (1.33 Å) identical to that of the (OH) anion. F is an

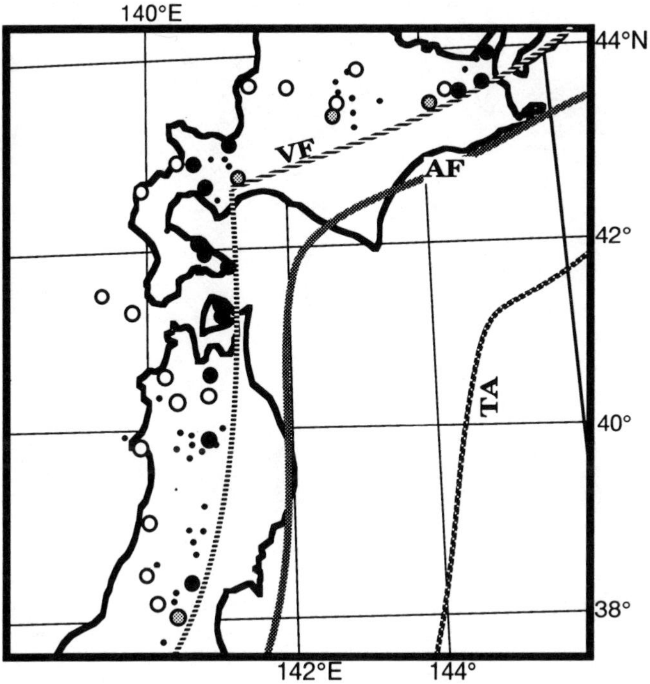

Figure 3.18 Phenocryst assemblages observed in volcanoes of the NE Japan arc. Closed circles—quartz without hornblende; open circles—hornblende without quartz; hatched circles—both quartz and hornblende phenocrysts. Small dots indicate other volcanoes. VF = Volcanic front; AF = aseismic front; TA = trench axis. *(Reproduced by permission from Sakuyama M, 1979.)*

exception among anions in that it is strongly partitioned into silicate melts rather than vapor phases (Koster van Groos and Wyllie, 1968), so that F remains in a magma during degassing or vapor-phase separation (Ishikawa et al., 1980). Concentrations of F in mafic lavas from the NE Japan arc (Ishikawa et al., 1980) support the occurrence of increasing H_2O contents in primary magmas toward the backarc side of this arc.

3.2.3 Element ratios

Ratios of certain incompatible elements are also observed to vary systematically in the across-arc direction within individual volcanic arcs. The ratios Rb/K, Rb/Sr, La/Ba, Nb/Zr, Th/U, Ba/Sr, and U/Pb

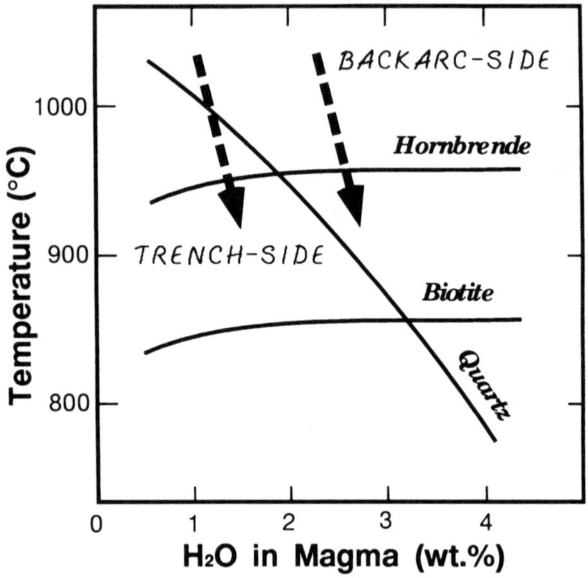

Figure 3.19 Schematic isobaric T-X_{H_2O} sections for andesite to dacite compositions. Liquidus temperatures of hornblende, biotite, and quartz and typical crystallization paths for trench- and backarc-side magmas are shown. *(Reproduced by permission from Sakuyama M, 1979.)*

usually increase at equivalent SiO_2 contents away from the volcanic front (Gill, 1981). These elements are all incompatible during melting of typical upper mantle mineral assemblages, and their partition coefficients are sufficiently similar that these particular element ratios can be expected to change little as a function of degree of partial melting (Figure 3.20). It follows that the variation observed in these element ratios requires either significant changes in across-arc source chemistry or the presence of different residual phases which may variably fractionate these elements. If a homogeneous mantle wedge source exists in the absence of subduction effects, then the across-arc variation in source mineralogy and/or chemistry may be related to the addition of subduction components. The identification of added subduction components and knowledge of the mechanism via which they are introduced into the mantle wedge are essential to our understanding of subduction zone magma genesis.

In order to assess how certain residual phases may govern the chemistry of a primary magma, the partition coefficients between

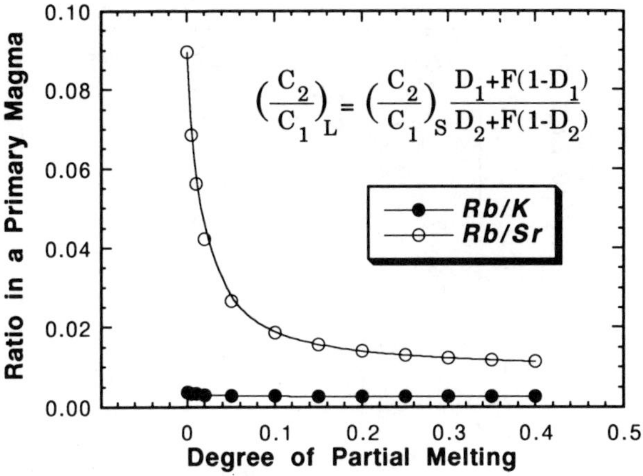

Figure 3.20 Variation in elemental ratios during partial melting of a typical lherzolite.

these phases and the silicate melts or fluids with which they coexist need to be known. In the absence of well-constrained partition coefficients, the identification of residual phases can be examined qualitatively using ratios of elements which have *different ionic radii but equivalent charge*. Matsui et al. (1977) demonstrated that the partitioning of elements with the same charge between minerals and silicate melts is governed largely by the ionic radius of the element and the crystal structure of the mineral. This was originally proposed by Onuma et al. (1968), who employed a partition coefficient *versus* ionic radius (PC-IR) diagram ("the Onuma diagram") to examine crystal structure controls on element partitioning. Figure 3.21 shows three simplified PC-IR diagrams for natural amphiboles and biotite phenocrysts in volcanic rocks. Several crystallographically nonequivalent positions for cations exist in most silicate minerals. On a PC-IR diagram, each position in a mineral shows up as a parabola-shaped peak for cations of a given valency (Onuma et al., 1968). In hornblende, for example, the following positions occur: T_1 and T_2 for small cations in tetrahedral coordination (e.g., Si^{4+}), the M_1, M_2, and M_3 positions for medium-sized cations in octahedral coordination (e.g., Mg^{2+} at ~70 pm), the slightly bigger M_4 position (~100 pm), and an additional large site (~140 pm) called the A position. A comparison of the different amphibole-melt pairs shows that, although the trends are similar,

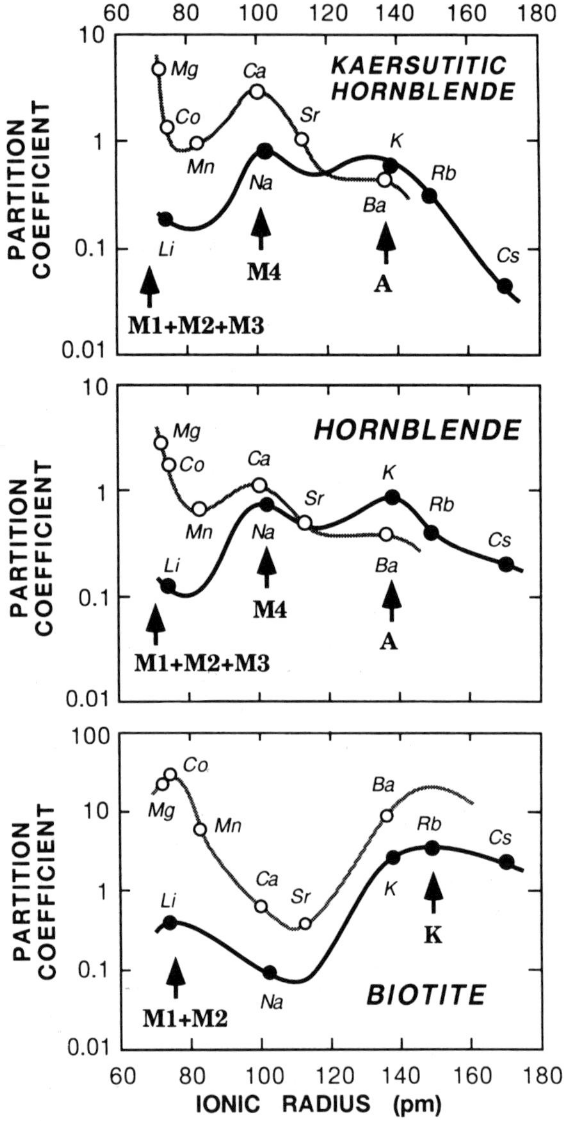

Figure 3.21 PC-IR (partition coefficient versus ionic radius) diagrams for amphiboles and biotite. *(Reproduced by permission from Matsui et al., 1977.)*

the absolute partition coefficients and the relative peak heights differ, indicating significant second-order crystal-composition/structure and possible physical controls such as temperature on trace element partitioning, as proposed by Irving (1978) and Banno and Matsui (1973). It follows that use of single-value partition coefficients, which have been derived for the same mineral but of different chemistry, may incorporate serious uncertainties. An example of trace element partitioning between black mica and silicate melt is also shown in Figure 3.21 and reveals the following five positions: T_1 and T_2 incorporate small cations with tetrahedral coordination, M_1 and M_2 include moderate-sized cations in sixfold coordination, and the K position. Note that it is Rb^{1+} not K^{1+} that is best accommodated in the K position.

The crystal structure control for element mineral-melt partitioning outlined above may also apply to fluid-mineral partitioning, as H_2O-rich fluids are likely to have an even more random and open structure than silicate melts. Two criteria need to be met if an element is to be used to identify phases which play key roles in producing the chemical heterogeneity in subduction zone magmas and their source due to mineral/fluid partitioning: (1) the element must be incompatible so that it is not strongly fractionated into phenocryst and residual phases during subsequent differentiation and partial melting processes, and (2) the element must be soluble and readily transported in H_2O-rich fluids under upper mantle *P-T* conditions, or else it may remain in the dehydrated material and not be transferred to the mantle wedge source. Among available elements with equivalent valence states, Rb-K and Ba-Sr-Pb are likely to satisfy these criteria. These elements have distribution coefficients generally smaller that 0.1 for major anhydrous mantle minerals; however, they have slightly higher coefficients for plagioclase, a potential fractionating phase in some basaltic magmas (D_{Rb} ~ 0.07, D_K ~ 0.1, D_{Ba} ~ 0.16, D_{Pb} ~0 .30; Gill, 1981; Meijer et al., 1990). Tatsumi et al. (1991) have shown that Rb/K and Ba/Pb ratios are little affected by small degrees of partial melting and possible plagioclase fractionation, with the Rb/K ratio varying little and the Ba/Pb potentially becoming smaller by up to ~20% compared to their primary melt ratios. These elements also have low ionic potentials (charge/ionic radius), so they are likely to behave as soluble ions in an aqueous fluid under upper mantle *P-T* conditions.

Rb/K and Ba/Pb ratios in lavas from three oceanic arcs, the Sangihe, Izu-Bonin, and Kurile, are higher in backarc-side lavas than trench-side lavas (Figure 3.22). If it is assumed that this results from a difference in the added components to their source regions, in the absence of preexisting across-arc compositional variation in the mantle wedge, then it follows that the probable hydrous fluid phase responsible for the variation in source chemistry has higher Ba/Pb and Rb/K ratios beneath the backarc side than the trench side. The residual mineral phases which may be responsible for these particular element fractionations will be evaluated in Chapter 4.

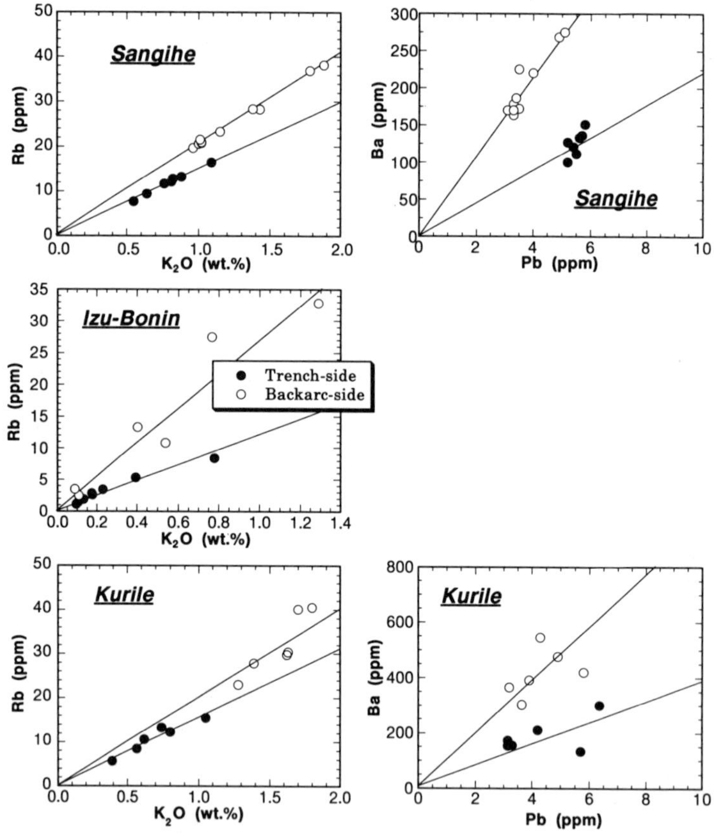

Figure 3.22 Rb/K and Ba/Pb ratio variation for basalts from the Sangihe (Tatsumi et al., 1991), Izu-Bonin (Tatsumi et al., 1992), and Kurile arcs (Bailey et al., 1989).

3.2.4 Radiogenic isotope ratios

There is no general rule for across-arc variation of Sr-Nd-isotopic ratios in subduction zone magmas; $^{87}Sr/^{86}Sr$ ratios decrease toward the backarc side in the Aleutian (Morris and Hart, 1983), Kurile (Zhuravlev et al., 1987), NE Japan (Hedge and Knight, 1969; Notsu, 1983), Izu-Bonin (Notsu et al., 1983), and Sangihe arcs (Jezek et al., 1979). $^{87}Sr/^{86}Sr$ values, on the other hand, are constant in New Britain (Page and Johnson, 1974; Woodhead and Johnson, 1993), Java (Whitford, 1975), and North Island, New Zealand (Gamble et al., 1993), and increase toward the backarc in the Sunda arc (Hutchison, 1982). Examples of these three types of variation are shown in Figure 3.23. A complex but particularly informative example occurs in the NE Japan arc (Nohda and Wasserburg, 1986; Nohda et al., 1988; Tatsumi et al., 1988) where backarc-side lavas display a temporal variation in both $^{87}Sr/^{86}Sr$ and $^{143}Nd/^{144}Nd$ ratios, with more depleted isotopic signatures occurring with time during the last 30 Ma, in contrast to trench-side lavas in which isotopic ratios have remained constant (Figure 3.24). In outline, $^{87}Sr/^{86}Sr$ ratios increase and $^{143}Nd/^{144}Nd$ ratios decrease toward the backarc side in Oligo-Miocene lavas, whereas the reverse pattern occurs for Quaternary lavas.

Across-arc variation in Sr-Nd-isotope ratios is probably best examined in oceanic arcs if geochemical fluxes from the subducted lithosphere are to be assessed and to minimize the potential effects posed by continental crustal contamination. In oceanic arcs, variation trends almost exclusively occur toward lower Sr- and Nd-isotopic ratios toward the backarc side (Figure 3.25), in contrast to the variation observed in continental arcs or arcs with thickened, mature crust, such as Indonesia. The Quaternary NE Japan arc is an exception in the context of the latter, being built on continental crust but with isotopic compositions becoming more depleted toward the backarc. This has been interpreted in terms of injection of more depleted asthenospheric mantle beneath the backarc-side volcanoes, in association with backarc basin development during Miocene time (Nohda et al., 1988; Tatsumi et al., 1989). By inference, the across-arc variation observed in other continental arcs may indicate greater contributions of more radiogenic isotopic continental materials (crust and/or subcontinental upper mantle) to backarc-side magmas.

The replacement of mantle wedge materials with isotopically de-

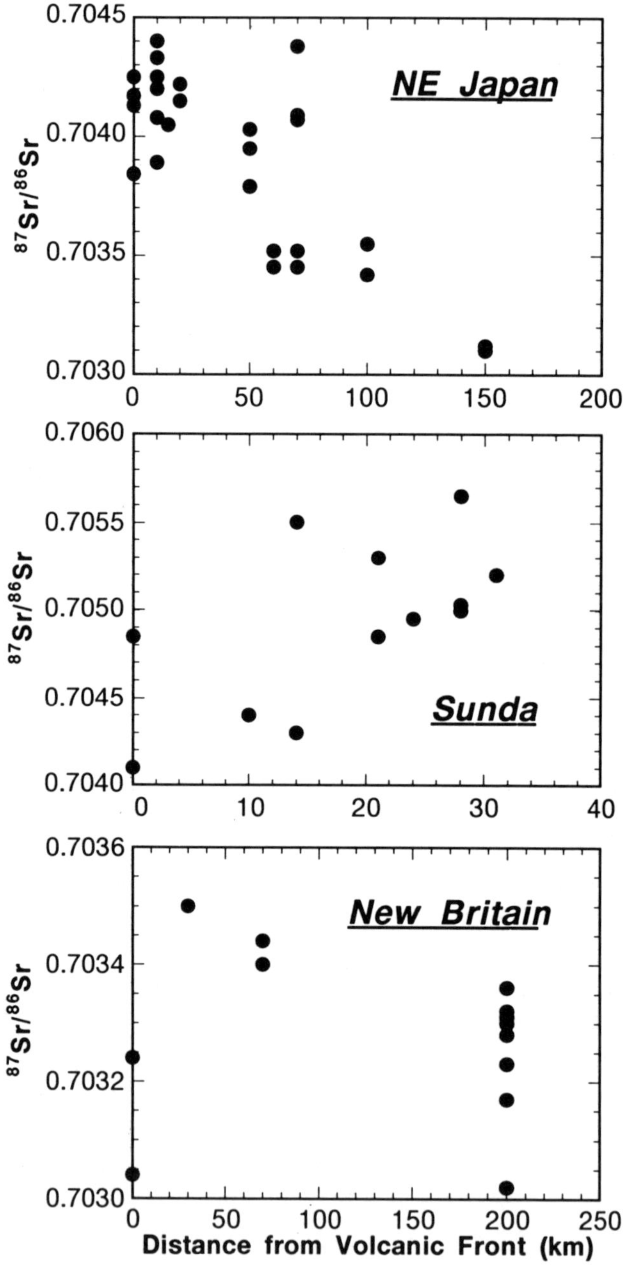

Figure 3.23 Across-arc variation in $^{87}Sr/^{86}Sr$ ratios for the NE Japan (Notsu, 1983), Sunda (Hutchinson, 1982), and New Britain arcs (Basaltic Volcanism Study Project, 1981).

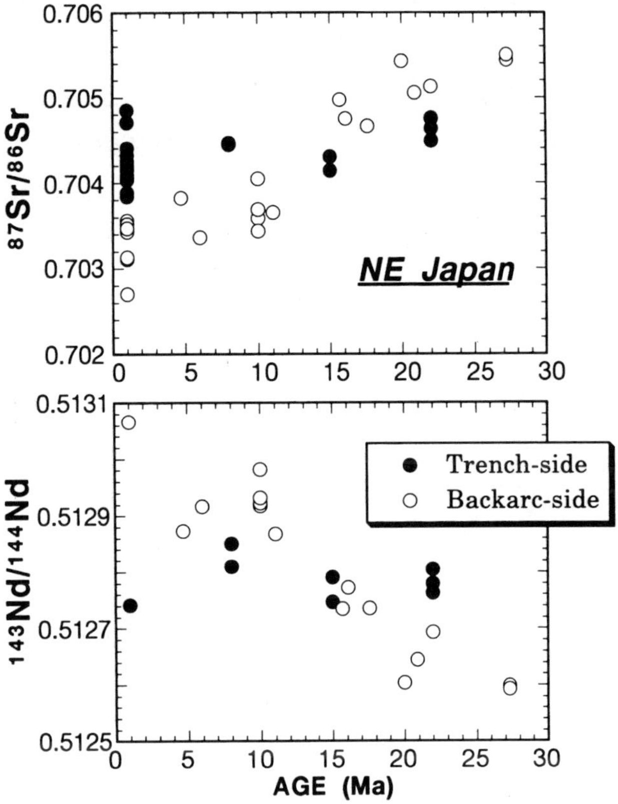

Figure 3.24 Secular variation in $^{87}Sr/^{86}Sr$ and $^{143}Nd/^{144}$ Nd ratios for the NE Japan arc. *(Data reproduced by permission from Nohda and Wasserburg, 1981, 1986; Nohda et al., 1988; Tatsumi et al., 1988.)*

pleted upper mantle material is clearly documented for lavas in the Neogene NE Japan arc. It follows therefrom that across-arc variation may be sensitive to both the presence of continental crust and the development of backarc basins. Across-arc radiogenic isotope variations arising directly from subduction-related processs are likely to be elucidated in oceanic arcs behind which backarc spreading has not taken place. The Aleutian and Sangihe arcs have this "ideal" setting and in both cases show typical oceanic-arc, across-arc variation trends in Sr-Nd-isotope ratios (Figure 3.25).

No obvious, generally applicable scheme of across-arc variation occurs for Pb isotopes. A decrease in $^{206}Pb/^{204}Pb$ ratios occurs toward the backarc side in the NE Japan (Hedge and Knight, 1969), New

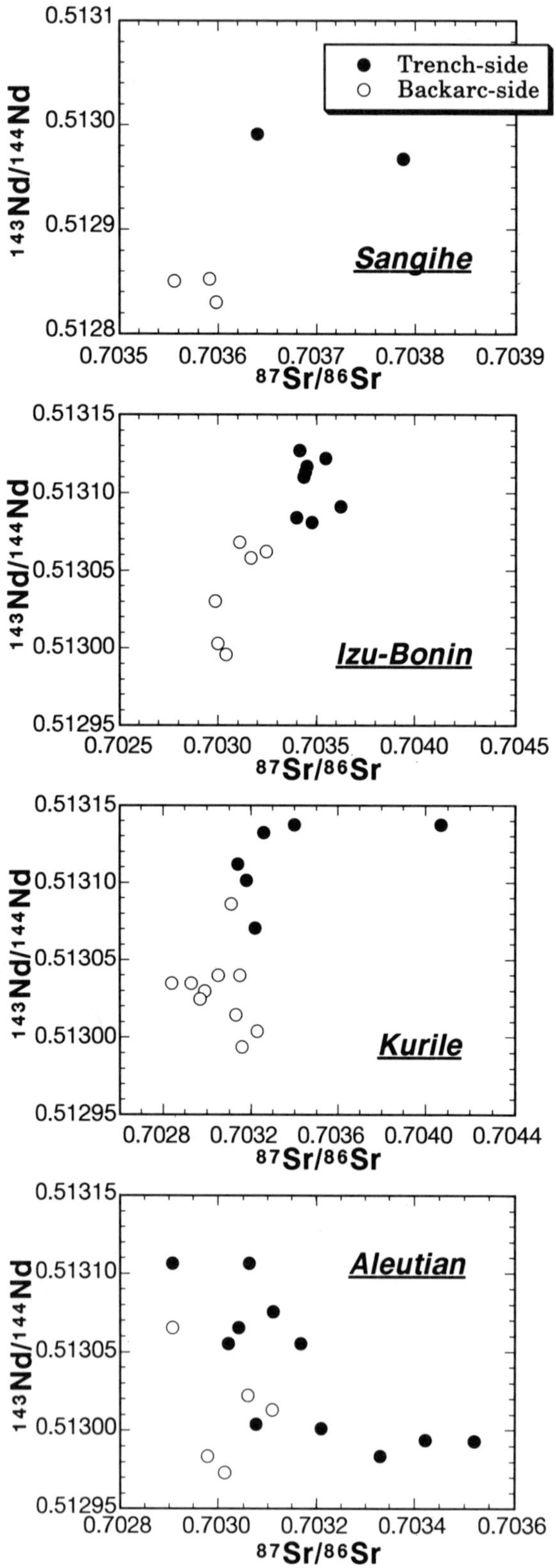
Trench-side
Backarc-side
Sangihe
Izu-Bonin
Kurile
Aleutian
$^{143}Nd/^{144}Nd$
$^{87}Sr/^{86}Sr$
0.5128
0.5129
0.5130
0.5131
0.7035
0.7036
0.7037
0.7038
0.7039
0.51295
0.51300
0.51305
0.51310
0.51315
0.7025
0.7030
0.7035
0.7040
0.7045
0.7028
0.7032
0.7036
0.7040
0.7044
0.7028
0.7030
0.7032
0.7034
0.7036

Zealand (Armstrong and Cooper, 1971), and New Britain arcs (Woodhead and Johnson, 1993), but the reverse variation occurs in the Sunda arc (Stolz et al., 1990). The former of these trends is consistent with a decrease in the subduction-derived influx of Pb away from the volcanic front, but data sets for lavas from oceanic arcs without backarc basins (e.g., Sangihe arc) may be required to shed further light on the origin of Pb-isotope systematics.

Figure 3.25 Across-arc variation in $^{87}Sr/^{86}Sr$ and ^{143}Nd ratios for lavas in several oceanic arcs. *(Data reproduced by permission from Tatsumi et al., 1991, 1992; Zhuravlev et al., 1987; Morris and Hart, 1983.)*

4

Hydrous Mineral Stabilities in Subduction Zones: Origin of Dual Volcanic Chains and Across-Arc Geochemical Variation

The characteristic spatial distribution of subduction zone volcanoes and evidence for characteristic across-arc geochemical variation have been outlined in preceding chapters. The formation of two volcanic chains at different and constant depths above the subducting lithosphere surface and the greater volcano density and magma volume erupted along the trench-side volcanic chain were demonstrated in Chapter 2. The systematic change in some incompatible element and isotopic ratios with distance from the volcanic front was outlined in Chapter 3. The origin of these characteristics will be examined in this chapter in the context of hydrous-phase stabilities in both the subducted oceanic lithosphere and the mantle wedge. It was suggested in Chapter 2 that the observed constant depths to the subducted lithopshere surface beneath both the volcanic front and backarc-side volcanic chains was due to pressure-dependent melt generation processes. We shall demonstrate that certain dehydration reactions are prime candidates for producing pressure-dependent melting phenomena in subduction zones.

4.1 Dehydration Reactions: Thermodynamic Background

The slope in pressure-temperature (*P-T*) space of a dehydration reaction is a function of the difference in entropy and volume of the reaction products and reactants as follows:

$$\left(\frac{dP}{dT}\right)_{\Delta G=0} = \frac{\Delta S}{\Delta V_s + \Delta V_f} = \frac{\Delta S}{\Delta V}$$

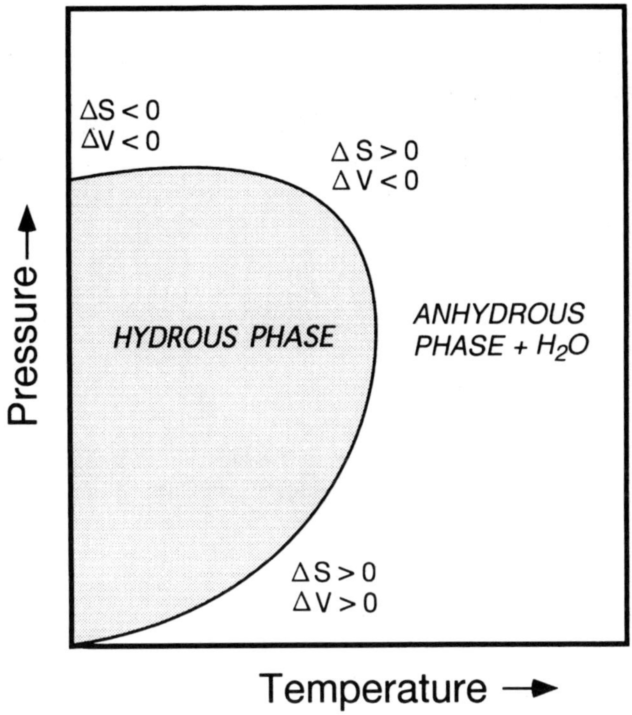

Figure 4.1 A schematic phase diagram illustrating the characteristic stability field shape for a hydrous phase.

where G = Gibbs free energy, T = temperature, S = entropy, V_s = volume of solid phases, V_f = volume of a fluid, and ΔV = volume change in the reaction.

Dehydration reactions are generally endothermic with positive ΔS, but ΔV changes greatly as a function of pressure due to the compressibility of H_2O. At relatively low pressures, the volume of H_2O is large and ΔV is positive; thus the slope of the reaction curve is positive (Figure 4.1). With increasing pressure and compression of H_2O, ΔV reduces and the slope becomes steeper. At a certain pressure, where $\Delta V = 0$, the P-T curve attains a maximum temperature, and beyond this pressure the slope will become negative, often forming a near-isobaric stability limit or ledge. In certain cases where ΔS also becomes negative at high pressure the slope may again become positive (Figure 4.1). Note that the high-pressure stability limits of hydrous phases are often isobaric reactions (Figure 4.1). These fundamental properties of dehydration reactions provide us with evidence for the breakdown of hydrous minerals playing a key role in the production of hydrous fluids and magmas in subduction zones.

The stability of a given hydrous phase in a complex natural rock system can be limited by the presence of coexisting phases, which are able to react with the hydrous phase. Consider the reaction involving the hydrous phase H:

$$H = A + H_2O$$

but in the presence of a phase B which will react with phase A to produce a third phase C. For the reaction A + B → C, $\Delta G_{AB \to C}$ is negative, and the following reaction should occur:

$$H + B = C + H_2O$$

This reaction requires that

$$\Delta G_{HB \to C} = \Delta G_{H \to A} + \Delta G_{AB \to C}$$

and $\Delta G_{HB \to C}$ must be smaller than $\Delta G_{H \to A}$, because $\Delta G_{AB \to C}$ is negative. Thus, a phase which is stable at certain *P-T* conditions may be not be stable there in a natural rock containing additional phases. A good example of this behavior is the stability of chlorite (clinochlore), $Mg_{10}Al_2Al_2Si_6O_{20}(OH)_{16}$, which is limited by the following reactions:

Clinochlore → 2 spinel + 2 forsterite + 2 enstatite + 8 H_2O
Clinochlore + 2 enstatite → 2 pyrope + forsterite + 8 H_2O
3 clinochlore + 5 muscovite
→ 8 kyanite + 5 phlogopite + quartz (coesite) + 12 H_2O

P-T curves for these reactions are shown in Figure 4.2 based on thermodynamic calculations by Delany and Helgeson (1978), Staudigel and Schreyer (1977), and Goto and Tatsumi (1990). The three reactions typify reactions occurring in pure chlorite, peridotite, and pelitic sediment systems, respectively. It is evident from Figure 4.2 that chlorite, which is stable at relatively high pressures in the pure system, will decompose at much lower pressures within both the upper mantle and subducted oceanic sediments.

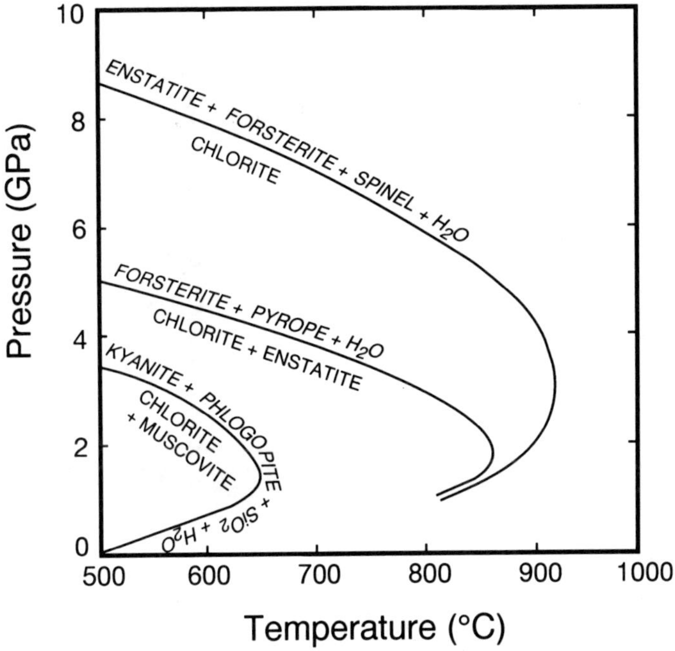

Figure 4.2 Stability limits of chlorite in the presence and absence of other reactable phases.

4.2 Dehydration Reactions in the Subducted Lithosphere

The subducted oceanic lithosphere comprises multiple layers of differing bulk composition. These layers are, from top to bottom, (1) a sediment layer of variable thickness but typically several hundred meters, (2) a basaltic crust approximately 7 km thick composed of pillow lavas, sheeted dykes, and gabbroic cumulates, and (3) a basal ultramafic layer comprising mantle peridotite of variable thickness, but typically 60–100 km. The bulk compositions of these respective layers determine the mineral phase assemblage present, including hydrous phases; thus it is necessary to examine the phase mineralogy and dehydration reactions of each layer separately.

Dehydration reactions occurring in sedimentary rocks have been extensively studied in high-pressure metamorphic belts. Metamorphosed sediments exhumed at the Earth's surface generally record reactions at relatively shallow depths (<~50 km) that may be attained in the accretionary wedge of some subduction zones but do not en-

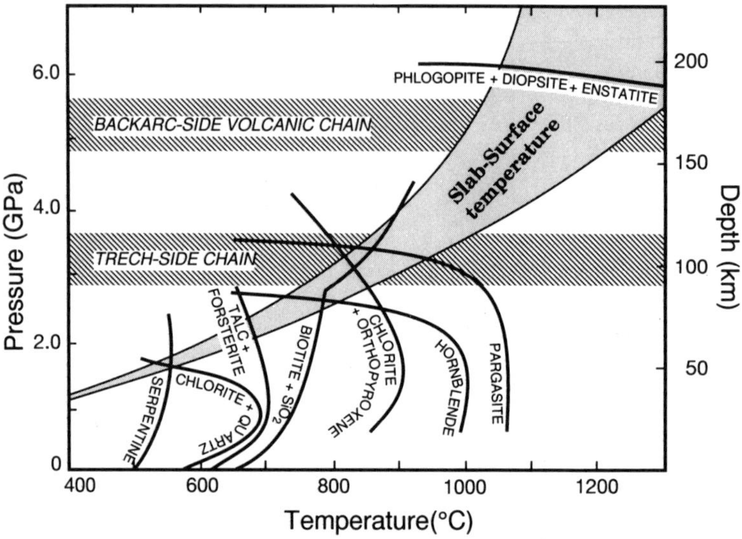

Figure 4.3 Stability limits of hydrous phases both in the subducted oceanic lithosphere and mantle wedge. Slab-surface temperature profiles as a function of pressure and temperature. *(Reproduced by permission from Toksöz et al., 1971 and Honda, 1985.)* These provide intermediate estimates among proposed profiles and subduction zone thermal structures.

compass the deeper levels to which sediments may be subducted. The occurrence of coesite, a high-pressure polymorph of quartz stable at pressures greater than 3.0 GPa, in Alpine (Chopin, 1984) and Caledonian localities (Smith, 1984) and in China (Wang et al., 1989) demonstrates that crustal materials can be subducted and have been subsequently exhumed from depths greater than 100 km in certain cases.

The composition of subducted sediment is variable; however, the most significant high-pressure dehydration reactions occurring in typical pelagic sediments (quartz-oversaturated) are likely to be

$$\text{Chlorite} + \text{quartz} \rightarrow \text{garnet} + H_2O$$

$$\text{Biotite} + \text{quartz} \rightarrow \text{orthopyroxene} + \text{K-feldspar} + H_2O$$

These reactions have been examined both experimentally and by thermodynamic calculation (e.g., Hsu, 1968; Delany and Helgeson, 1978; Bohlen et al., 1983; Banno et al., 1986), the results of which are shown in Figure 4.3 for chlorite and biotite with Mg/Fe mol ratios of unity. The stability of these phases in natural samples, which have

Mg/Fe ratios less than unity, will be limited to lower temperatures. Thus, if temperatures exceed 700°C along a moderate subduction geotherm (Figure 4.3), no hydrous phase may be present in the sediment layer of the subducted oceanic crust at depths greater than 70 km (2.5 GPa). However, the occurrence of phengite, $K_2(Mg,Fe)Al_3Si_7AlO_{20}(OH)_4$, in some coesite-bearing rocks (Chopin et al., 1991) indicates that hydrous phases in subducted sediments may transport H_2O to deeper levels and possibly to beneath the volcanic arc. Experimental results show that phengite stability at high pressures (>3.0 GPa) is limited to relatively low temperature conditions (Massonne and Schreyer, 1989). Furthermore, the reaction phengite + talc → phlogopite + quartz + kyanite + H_2O may determine the upper stability limit of phengite, such that the stability of phlogopite (biotite) may then govern the highest-pressure limit of H_2O transport in the subducted sediment layer (Miyashiro, 1973; Wang and Banno, 1987).

Dehydration reactions in basalt compositions may be more variable than in sediments, since (1) basalts are not saturated with quartz and (2) amphiboles spanning a wide compositional range are the principal hydrous phases. The important dehydration reactions which take place at relatively high pressures in the subducting basaltic crust are

$$\text{Chlorite} + (\text{albite, actinolite, epidote}) \rightarrow \text{amphibole} + H_2O$$

$$\text{Amphibole} \rightarrow \text{clinopyroxene} + \text{garnet} + H_2O$$

It should be stressed that the amphibole which participates in these reactions is hornblende, not pargasite or other end-member amphiboles in the basaltic system (e.g., Gilbert et al., 1982). The hornblende-decomposition reaction curve for a low-K tholeiite basalt under conditions of $P_{H_2O} = P_{total}$ is drawn after Lambert and Wyllie (1972) in Figure 4.3. The high-pressure-stability limit of hornblende has a negative slope, and this may contract with lower P_{H_2O} (Allen and Boettcher, 1978). Therefore, the pressure limit of hornblende and of H_2O transport in the subducting basaltic crust is unlikely to exceed a depth of 80 km (~2.5 GPa).

It has been suggested that phengite (muscovite) may be a significant hydrous phase in the subducted basalt beyond the stability limit of hornblende (Nicholls et al., 1980). However, phengite/muscovite is unlikely to be present at such depths, since it will decompose within the amphibolite facies by reaction with amphibole (Miyashiro, 1973). A possible phase carrying H_2O to deeper levels is an epidote-group mineral that has been documented in a diamond/coesite-bearing eclogite from China (Maruyama, personal communication). Under normal metamorphic conditions, however, epidote decomposes to produce An-rich plagioclase and may be absent in the amphibolite facies (Miyashiro, 1973). Further experimental work on epidote stability is required to ascertain its potential significance to subduction zone processes. Lawsonite has also been demonstrated to be stable to pressures of 10 GPa (Schmidt and Poli, 1994) and is a candidate for transporting H_2O to deeper levels, but for this reason is unlikely to contribute to the production of subduction zone magmas.

Ultramafic rocks sampled from the ocean basins are commonly serpentinized, indicating the likely presence of serpentine in the subducted peridotite layer of the oceanic lithosphere. Ringwood (1975) suggested that dehydration of serpentinite and derivative, dense hydrated magnesian phases (see below) might be responsible for partial melting of the subducted oceanic lithosphere at depths up to 300 km. Serpentine undergoes the dehydration reaction series

$$\text{Serpentine} \rightarrow \text{olivine} + \text{talc} + H_2O$$

$$\text{Olivine} + \text{talc} \rightarrow \text{orthopyroxene} + H_2O$$

Both reactions are approximately isothermal at pressures less than 4 GPa (Figure 4.3); the former takes place at ~600°C and the latter at ~700°C (Kitahara et al., 1966). These reactions are likely to take place beneath the forearc under normal subduction zone geotherms, and are inconsistent with the depths of the pressure-dependent reactions which are implicated in the generation of subduction zone magmas (Figure 4.3). However, under extremely low temperature conditions serpentine can be stable to much deeper levels (Ringwood, 1990).

In ultramafic rock compositions, chlorite invariably crystallizes during hydration/alteration processes and may be one of the principal hydrous phases in the peridotite layer within the oceanic lithosphere. The stability of chlorite in a peridotite system is governed by the following reaction (Goto and Tatsumi, 1990):

$$\text{Chlorite} + \text{orthopyroxene} \rightarrow \text{garnet} + \text{olivine} + H_2O$$

Results of thermodynamic calculations by Goto and Tatsumi (1990) for this reaction are shown in Figure 4.3. They illustrate that chlorite

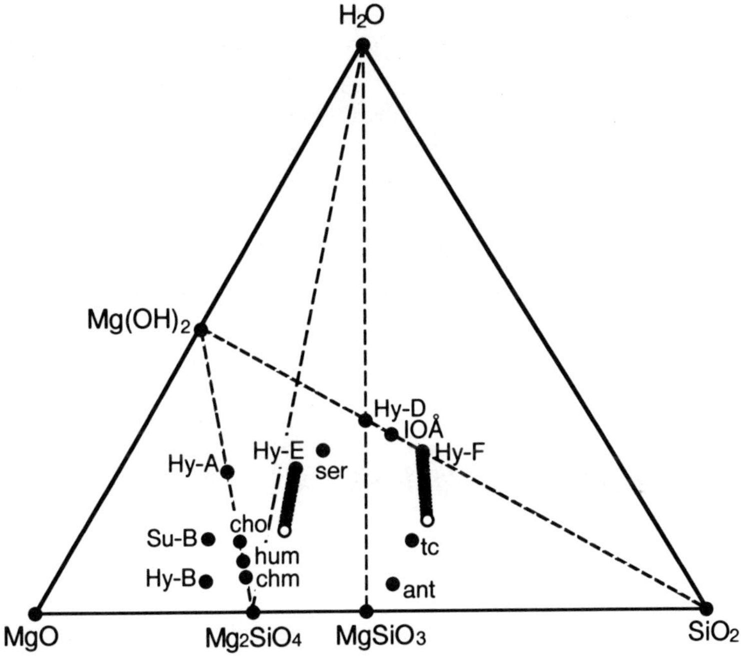

Figure 4.4 Selected phases in the system MgO-SiO_2-H_2O. Hy = hydrous; Su = Superhydrous; cho = chondrodite; hum = humite; chm = clinohumite; ser = serpentive; 10Å = 10Å-phase; tc = talc; ant = anthophyllite. *(Reproduced by permission from Thompson, 1992.)*

decomposes via a partially pressure-dependent reaction, in the vicinity of 3.0–3.5 GPa in the subducted oceanic lithosphere and therefore may play a role in magma production beneath the trench-side volcanic chain.

Dense hydrated magnesian silicates (DHMSs) have been synthesized in high-pressure experiments (Figure 4.4) and are potential candidates for transporting H_2O to much deeper levels within subduction zones. Crystallization of DHMSs with MgO/SiO_2 mol ratios greater than 2 is limited to mantle peridotites with high MgO/SiO_2 ratios, such that these phases might form in the subducting lithosphere following the extraction of SiO_2 by hydrous fluids released during serpentine dehydration (Ringwood, 1977). On the other hand, DHMSs with MgO/SiO_2 ratios less than 2 can crystallize in normal peridotite compositions. Yamamoto and Akimoto (1977) investigated the MgO-SiO_2-H_2O system up to 7.7 GPa and 1225°C and proposed that the 10-Å phase ($Mg_3Si_4O_{10}(OH)_2 2H_2O$), a phyllosilicate with a *c*-lattice constant near 10 Å, may form by dehydration of talc at pressures above 3.0 GPa.

A recent experimental study by Wunder and Schreyer (1992), however, shows that the 10-Å phase is metastable and does not have a role as a natural phase in subduction zones or the mantle elsewhere. Other DHMSs identified by alphabetic abbreviations (phases A, B, D, and E; Figure 4.4) are stable at pressures greater than 6 GPa under normal subduction zone conditions and cannot be considered to contribute to magma production beneath volcanic arcs.

This outline of hydrous phase stabilities in the subducted lithosphere supports the suggestion by Anderson et al. (1976) and Tatsumi (1989) that dehydration of the subducted lithosphere is likely to be largely completed at shallow levels beneath the forearc region and therefore does not play a direct role in causing partial melting of the overlying mantle wedge.

4.3 Hydrous Phase Stability in the Mantle Wedge and Formation of Dual Volcanic Chains

4.3.1 Dehydration reactions

Hydrous phases present in the subducted oceanic lithosphere largely decompose to release H_2O as a supercritical fluid at depths <100 km, corresponding to those beneath the forearc. This fluid is buoyant, so the H_2O released will migrate upward into the overlying mantle wedge, where it can react with peridotite minerals and recrystallize hydrous phases. The resultant hydrated peridotite is then capable of being transported downward along the top of the subducting plate due to viscous coupling between the rigid oceanic lithosphere and the mantle wedge (e.g., McKenzie, 1969; Andrews and Sleep, 1974).

The hydrous mineral phases in the hydrous peridotite layer at the base of the mantle wedge will be equivalent to those in the peridotite layer of the subducted oceanic lithosphere, but may also include the presence of K-bearing phases such as amphibole and phlogopite. Although mantle peridotites are generally depleted in potassium and other large-ion lithophile elements, enrichment in these elements may occur via aqueous fluids migrating from the subducted lithosphere (see also Chapter 6) and K-bearing phases crystallized accordingly in the hydrated parts of the mantle wedge. As hydrated mantle wedge material is dragged to higher pressure and temperature, the hydrous mineral assemblage may undergo the following dehydration reactions (see also Figure 4.5):

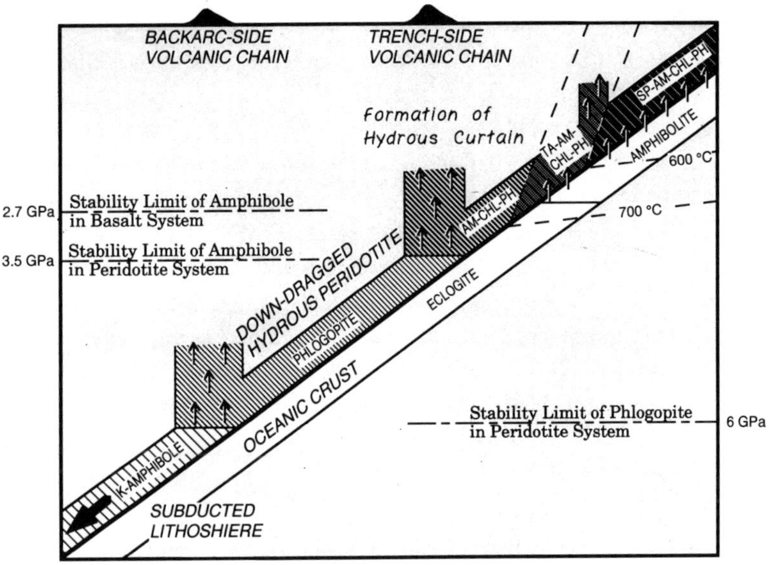

Figure 4.5 A schematic model of dehydration reactions occurring in the downdragged hydrous peridotite layer and associated formation of hydrous curtains rising into the mantle wedge. Note that decomposition of hydrous phases in the subducted oceanic crust is largely completed beneath the forearc region, and that pressure-dependent dehydration of amphibole/chlorite and phlogopite take place in the hydrous layer immediately beneath the trench- and backarc-side volcanic chains, respectively. SP = serpentine, AM = amphibole, CHL = chlorite, PH = phlogopite, TA = talc.

Serpentine → olivine + talc + H_2O

Talc + olivine → orthopyroxene + H_2O

Chlorite + orthopyroxene → olivine + garnet + H_2O

Amphibole → clinopyroxene + garnet + H_2O

(K-bearing amphibole + orthopyroxene
→ phlogopite + clinopyroxene + olivine + garnet + H_2O)

Phlogopite + orthopyroxene + clinopyroxene
→ K-amphibole + garnet + olivine + K_2O + H_2O

The *P-T* curves for these reactions are shown in Figure 4.3. Note that the amphibole present in the peridotite system (i.e., pargasite) is more stable at higher temperatures and pressures than is the amphibole in a basalt system (i.e., hornblende), due to the higher Mg/Fe ratio of pargasite compared with hornblende. Moreover, the predicted depth of breakdown of both pargasitic amphibole and chlorite, based on the intersection of subducting plate geotherms and the pressure-dependent parts of their reaction curves, matches closely the depth to the surface of the subducted plate surface beneath the trench-side volcanic chain. This constitutes compelling evidence for the generation of magmas beneath the trench-side volcanic chain as a result of the *breakdown of these phases in the downdragged hydrous peridotite layer,* as opposed to dehydration reactions in the subducted oceanic lithosphere.

Can other dehydration reactions explain the formation of the backarc-side volcanic chain? High-pressure experiments by Sudo and Tatusmi (1990) indicate that the decomposition of phlogopite takes place at ~6 GPa in peridotite compositions, a pressure which corresponds approximately with the depth to the top of the subducted lithosphere beneath the backarc-side volcanic chain. The chemical formula of the very high pressure amphibole formed as a consequence of phlogopite dehydration is $K_{1.9}Ca_{1.1}Mg_5Si_{7.9}Al_{0.1}O_{22}(OH)_2$, and similar to that of potassium richterite (Sudo and Tatsumi, 1990). Experimental constraints on the stability of this amphibole composition are not yet complete (Foley, 1991).

4.3.2 Formation of hydrous curtains

The series of dehydration reactions which take place at different depths within the hydrous peridotite layer, at the base of the mantle wedge, will result in the formation of three regions of ascending H_2O-rich fluid into the overlying mantle wedge. We shall refer to these regions as hydrous curtains. These hydrous curtains are instrumental to the initiation of melting as they ascend into hotter regions of the mantle wedge, and their positioning in the mantle wedge can be related to the spatial distribution of volcanism at the surface.

Serpentine in the hydrated parts of the mantle wedge is produced by reaction of subducting plate-derived H_2O with olivine and orthopyroxene, the major constituents of mantle peridotites. Serpentine-like chlorite contains a large amount of H_2O (~13 wt%), but the abundance of chlorite is restricted by the amount of preexisting aluminous phases. The assemblage talc + chlorite + amphibole + phlogopite (see Figure 4.5) can therefore accommodate less H_2O than serpentinized peridotites. A simple calculation of the normative composition of a hydrous peridotite in the system CaO-MgO-Al_2O_3-SiO_2 provides a guide to the change in H_2O content that will occur with

Table 4.1 Modal compositions and H_2O contents of hydrous peridotites

ANHYDROUS			<600°C			<700°C			<3.5 GPa		
Olivine	44.6	50.0	Serpentine	68.0	51.9	Talc	21.6	16.6	Chlorite	21.7	—
Orthopyroxene	38.0	22.4	Chlorite	19.9	—	Chlorite	21.5	—	Amphibole	—	16.0
Clinopyroxene	11.8	11.8	Amphibole	—	14.8	Amphibole	—	15.9	Olivine	38.0	56.8
Spinel	5.6	—	Orthopyroxene	1.6	28.7	Olivine	45.6	62.5	Orthopyroxene	28.8	22.2
Garnet	—	15.8	Clinopyroxene	10.4	4.6	Clinopyroxene	11.3	5.0	Clinopyroxene	11.5	5.0
			H_2O	11.4	7.1	H_2O	3.9	1.2	H_2O	2.8	0.4

Calculations are based on the system CaO-MgO-Al_2O_3-SiO_2 (CMAS). Anhydrous peridotite compositions are close to that of pyrorite (Green and Ringwood, 1967). In this system, amphibole is represented by tschermakite. The relative amount of amphibole and chlorite varies probably as a function of pressure and temperature, so that two extreme cases in which chlorite or amphibole is the only Al_2O_3-bearing phase are presented.

progressive dehydration reactions in the downdragged hydrous peridotite layer (Table 4.1) and demonstrates that a comparatively large amount of H_2O is released by decomposition of serpentine. This fluid will migrate upward as a hydrous curtain, comprising a hydrous, serpentine-free mineral assemblage, which ascends into the mantle wedge above the 600°C isotherm. The intersection of the subducting plate surface geotherm with the *P-T* curve defining the stability limit of serpentine (Figure 4.3) determines the source and location of the hydrous curtain. Except for cases with extremely cold geotherms, the hydrous curtain will form beneath the forearc mantle wedge where it is unlikely to exceed the peridotite solidus and initiate melting, as shown in Figure 4.5. However, the occurrence of unusual near-trench volcanism involving high-Mg andesites in some subduction zones (see Chapters 1 and 2) may reflect anomalously high temperatures in the shallow forearc mantle wedge and may be a consequence of partial melting of the serpentine-dehydration hydrous curtain (Tatsumi and Maruyama, 1989).

Talc in the hydrous peridotite layer will decompose at ~700°C (Figures 4.3 and 4.5) in the presence of chlorite and amphibole. The amount of H_2O held in chlorite-amphibole peridotites (2.8–0.4 wt%) is roughly equivalent to that in talc-bearing peridotites (3.9–1.2 wt%) (Table 4.1), and thus a hydrous curtain is unlikely to emanate from the decomposition of talc.

Amphibole (K-bearing pargasite) and chlorite will decompose at depths of ~110 km in the downdragged hydrous peridotite layer, corresponding to that beneath the volcanic front. Phlogopite may be the only hydrous phase to progress to deeper levels. The amount of phlogopite in the hydrous peridotite layer is likely to be determined largely by the amount of potassium added via subducting slab-derived fluid phases, a quantity which is difficult to estimate. If all the H_2O released from chlorite (0.4–2.8 wt%) and amphibole dehydration were to be held in phlogopite, the K_2O content of the peridotite layer would need to be in the range 1–7 wt%. This is improbable; therefore the change in hydrous mineral assemblage from amphibole-chlorite to phlogopite will result in the formation of a second hydrous curtain at a depth of ~110 km, directly beneath the volcanic front. Thereafter, the breakdown of phlogopite and crystallization of K-amphibole at ~6 GPa may form a third hydrous curtain at a level corresponding to that beneath the backarc-side volcanic chain.

In reality the reactions occurring within both the subducted lithosphere and the downdragged hydrous layer are likely to be more complex than we have summarized, due to solid solution effects and the additional presence of minor hydrous phases such as phengite and chloritoid, which may transport and release H_2O to and at different depths. These complexities will serve to disperse both the release of H_2O into and the generation of melt within the mantle wedge, and may account for the dispersion occurring about the observed mean depths to the surface of the subducting plate beneath the volcanic front and backarc-side volcanic chains. However, the consistency between the mean depths to the subducted plate surface beneath these volcanic chains and those of the decomposition of amphibole/chlorite- and phlogopite-peridotite provide compelling evidence for the dominant role which these particular dehydration reactions play in the formation of subduction zone magmas.

4.4 Variation in Amount of Released H_2O and Across-Arc Density of Volcanoes

Several lines of evidence, which were reviewed in Chapter 2, suggest the localized distribution of partially molten regions within the mantle wedge, including (1) tomographic definition of a low-V, probably high-T, column beneath active volcanoes in NE Japan, (2) shear-wave splitting evidence for partially molten mantle regions beneath central Japan, and (3) the distinct incompatible element characteristics observed between NE Japan volcanoes separated by as little as 10 km. Localized regions of partial melting may be related to Rayleigh-Taylor gravitational instabilities, or rising diapirs of partially molten material, within the mantle wedge. These instabilities will develop where a buoyant, less dense, fluid underlies a denser fluid, as may be the case where partially molten peridotite or partial melt itself is overlain by subsolidus mantle peridotite. If an instability in partially molten mantle material forms a rising bulb, it is referred to as a mantle diapir. The development of Rayleigh-Taylor instabilities has a characteristic wavelength (d) which can be related to certain physical properties of the system (after Selig, 1965); that is;

$$d \propto h\left(\frac{\eta_1}{\eta_2}\right)^{\frac{1}{3}}$$

where h is the thickness of the less dense fluid and η_1 and η_2 are the viscosities of the upper and lower layers ($\eta_1 >> \eta_2$). If the viscosities of both mantle materials and partial melts are the same beneath the trench- and backarc-side volcanic chains, then a possible cause of greater volcano numbers in the former could be a thinner layer of partially molten mantle beneath the volcanic front.

Another mechanism which may account for the difference in volcano numbers has been proposed by Marsh (1979), who examined the spacing of volcanic centers based on a set of fluid instability and numerical experiments. Marsh showed that if a layer of partially molten peridotite dips at the angle of subduction, then a greater number of instabilities will develop at the upper corner of the layer, a position which may be equated with that of the trench-side volcanic chain. This mechanism can also explain the inverse correlation between the angle of subduction and the width of the volcanic zone or the distance between two volcanic chains (Marsh, 1979), a proposal which has received favor from several authors (Sakuyama, 1983b; Kushiro, 1983; Tatsumi et al., 1983). The formation of a continuous dipping layer of partially molten mantle beneath the volcanic arc may require melting to be largely a function of temperature, which, given the large variation in mantle wedge thermal structure expected for different subduction zones of contrasting maturity and subducting lithosphere ages (e.g., Peacock, 1990), is inconsistent with the apparent pressure-related onset of melting at ~110 km beneath the volcanic front.

It seems equally plausible that the greater density of volcanoes and volume of volcanism at the trench-side chain may be attributed to a greater rate of melt production there. We propose that this stems from a likely larger fluid flux into the mantle wedge beneath the trench-side rather than the backarc-side chain, and we outline a simple mechanism to explain the observed across-arc distribution of volcanism in this context. This can be related simply to the derivation of a greater amount of H_2O from the breakdown of pargasitic amphibole and chlorite beneath the trench-side magma source region compared with the decomposition of phlogopite beneath the backarc-side source region. However, this in itself does not explain why a greater H_2O influx results in a greater density of volcanoes? As outlined, volcano spacing may be related to the localization of partially molten material in the mantle wedge in the form of Rayleigh-Taylor instabilities. The

instablity wavelength of a partially molten region is unlikely to be influenced directly by the relative amount of H_2O influx, but rather will be a function of the thickness of the partially molten region and its relative viscosity to that of the overlying unmolten mantle wedge. Alternatively, we propose that the H_2O flux may determine the time to produce a rising partially molten instability, and we illustrate this using a simple bulb or diapir model, following the equations outlined by Marsh (1979).

The ascent velocity of a fully developed spherical bulb, V, is given with the Hadamard-Rybsznski extensions of Stokes' law to fluid bodies, by

$$V = \frac{1}{3}\frac{R^2 g\,\Delta\rho}{\eta_1}\left(\frac{\eta_1 + \eta_2}{\eta_1 + \frac{3}{2}\eta_2}\right)$$

where R = radius of the bulb, g = gravitational accerelation, $\Delta\rho$ = density contrast, and η_1 and η_2 are, respectively, the viscosity of the overlying unmolten peridotite and the partially molten peridotite. The term in parentheses applies to mantle wedge conditions and assumes a value between unity ($\eta_1 >> \eta_2$) and 4/5 ($\eta_1 = \eta_2$). Given a bulb growing from a constant influx of partial melt and that this flux is a function of the aqueous fluid flux (Q) to the partially molten region, the bulb will separate and rise when its radius enlarges to the point where the ascent velocity is greater than the rate of radius growth:

$$\frac{dR}{dt} = \frac{Q}{4\pi R^2}$$

The critical radius (R_0) can be obtained by equating these equations:

$$R_0 = \left(\alpha\frac{Q\eta_1}{\pi\,\Delta\rho\,g}\right)^{\frac{1}{4}}$$

for $1 > \alpha > 4/5$, and the time taken to reach the critical radius (t_b) is

$$t_b = \left(\frac{\frac{4}{3}\pi R^3}{Q}\right) \propto \left(\frac{1}{Q}\right)^{\frac{1}{4}}\left(\frac{\eta_1}{\Delta\rho}\right)^{\frac{3}{4}}$$

If the density contrast between partially molten and unmolten mantle material and the viscosity of unmolten mantle materials are compa-

rable beneath the trench- and backarc-side volcanic chains, and if each bulb is the source of an individual volcano within the overlying volcanic arc (Sakuyama, 1983b), then a greater influx of fluid derived from the breakdown of amphibole and chlorite could be equated with a greater number of volcanoes at the trench-side volcanic chain.

4.5 Across-Arc Variation in Elemental and Isotope Ratios—Due to Difference in Fluid Composition?

The pressure-dependent dehydration reactions occurring at 3.5 and 6 GPa in downdragged hydrous peridotite layer release aqueous fluid phases which are not pure H_2O but must contain additional dissolved components. The concentration of other elements within these fluids will be a function of partitioning between the fluid and the solid phases present in each case. It was argued in Chapter 3 that this partitioning behavior is likely to be controlled in part by the crystal structure of the solid phases (Section 3.2.3), and it can be asserted therefrom that the composition of the H_2O-rich fluid phases released beneath the trench- and backarc-side volcanic chains may be distinct. This logic, if correct, has significant ramifications for the origin of certain across-arc geochemical variations.

It has been observed (Section 3.2.3) that Rb/K and Ba/Pb ratios in the magma source region are higher beneath backarc-side than volcanic front volcanoes. If the mantle wedge is homogeneous in chemical composition in the absence of subduction effects, then the across-arc variation in these ratios must result from differences in the added fluid phases (subduction-derived components). Given crystal structure control of element partitioning between solid and melt or fluid phases, the production of a fluid phase with lower Rb/K and Ba/Pb ratios may be attributed to the presence of phlogopite beneath the trench-side volcanic chain and potassic-amphibole beneath the backarc-side volcanic chain. This arises from the decomposition of pargasitic amphibole and formation of phlogopite within the downdragged hydrous peridotite layer beneath the trench-side volcanic chain, and from the breakdown of phlogopite to give K-rich-amphibole beneath the backarc-side chain. Phlogopite is known to preferentially include Rb over K, whereas the opposite case exists for amphibole.

Jakes and White (1970) observed that the tholeiitic rocks erupted from trench-side volcanoes have lower Rb/K than alkaline rocks erupted from backarc-side volcanoes. They attributed this to the

involvement of different proportions of amphibole and phlogopite during decomposition or melting of the subducted oceanic lithosphere. This interpretation was based on the occurrence of lower Rb/K in amphibole than in phlogopite (biotite). The conclusion of Jakes and White (1970) is similar to the one presented here, except that we emphasize the importance of fluid equilibration with the residual phases formed by dehydration reactions rather than the element ratios of the dehydrating phases themselves.

Across-arc variation in Sr-Nd-isotope ratios, with lower $^{87}Sr/^{86}Sr$ and $^{143}Nd/^{144}Nd$ occurring in the backarc-side lavas and hence also in their magma source region, can be explained by either of two simple mixing models (Figure 4.6). Given the process of fluid-phase migration during subduction outlined above, the isotopic composition of both the downdragged hydrous peridotites and the fluid phase released therefrom through decomposition of hydrous phases are required to lie on a mixing line between the composition of the mantle wedge and the subduction-derived fluid. The composition of subduction-derived fluids can only be evaluated qualitatively, because it is difficult to establish the Sr/Nd ratios of either altered MORB- or sediment-derived fluid phases which enter the mantle wedge. In one model (A in Figure 4.6), it has been assumed that the mantle wedge has an isotopic composition equivalent to that of the N-type MORB source, which is commonly believed to be resident in the mantle wedge (e.g., Perfit et al., 1980; Ellam and Hawkesworth, 1988), and that the subduction-derived fluid has a major contribution from both subducted sediment and altered oceanic crust. In this model the observed across-arc variation in Sr-Nd isotopes could be explained if a Sr/Nd ratio attained by the hydrous-peridotite-derived fluid released beneath the volcanic front is higher than that released beneath the backarc-side volcanic chain (e.g., DePaolo and Wasserburg, 1979). This possibility is consistent with the higher Sr/Nd values observed in lower K, and hence volcanic front magmas (Chapter 3), given the reasonable assumption that these characteristics are attributed to the added fluid phase. However, a number of observations might be used to bring the validity of this model into question, including (1) Sr/Nd ratios in amphiboles overlap or are smaller than those in phlogopite (10–100 versus >75, respectively; e.g., Basu and Tatsumoto, 1980; Menzies and Murthy, 1980); and (2) amphibole contains far greater concentrations of Sr and Nd than phlogopite

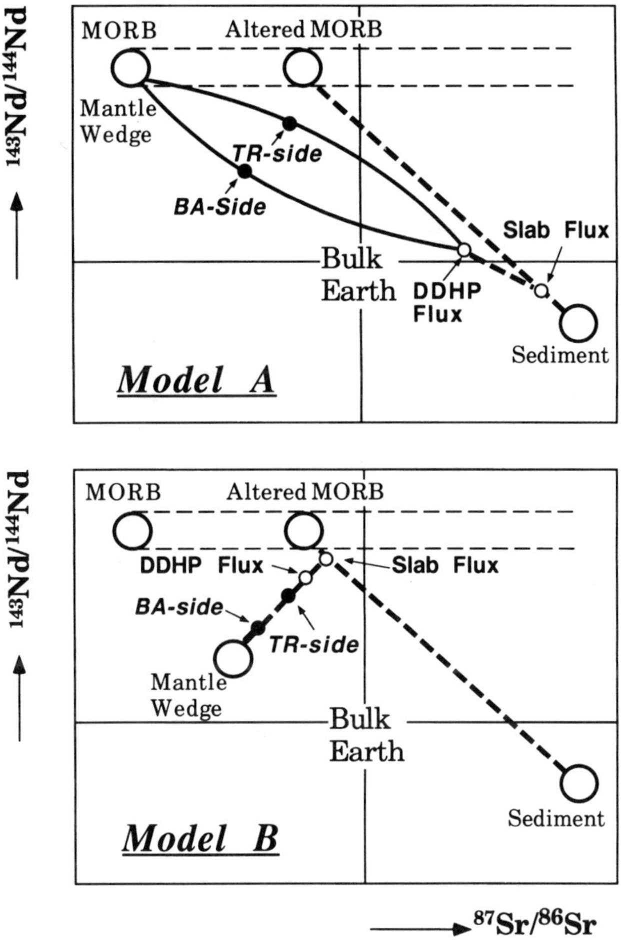

Figure 4.6 Two possible models which are consistent with across-arc variation in Sr- and Nd-isotope ratios. TR = trench; BA = backarc; DDHP flux = fluid from downdragged hydrous peridotite layer. *(Reproduced by permission from Tatsumi et al., 1992.)*

(300–1000 versus 150 ppm for Sr and 10–50 versus <2 ppm for Nd respectively). These observations suggest that greater amounts of Sr and Nd may be released by amphibole dehydration than by phlogopite dehydration and, accordingly, that the trench-side magmas might more closely reflect the composition of the hydrous-peridotite-derived fluid (Figure 4.6).

An alternative model (B in Figure 4.6) may also explain the

observed across-arc isotope variation systematics. If the altered-MORB crust is assumed to dominate the composition of the slab-derived fluid, *at least in terms of Sr-Nd isotopes* (this is less reasonable for Pb isotopes as subducted sediment is by far the dominant reservoir of Pb in the subducted lithosphere), then the isotopic composition of the mantle wedge cannot be matched with the MORB source and must plot between the MORB source and the bulk earth, within the mantle array (Figure 4.6). This position is similar to some enriched ocean island basalt (OIB) mantle sources, which have been proposed as possible mantle wedge components (e.g., Stern, 1981; Morris and Hart, 1983). Across-arc variation in Sr- and Nd-isotopic compositions may therefrom be reasonably well understood as a consequence of two-component mixing between a dominantly altered MORB-derived fluid and the mantle wedge. This is also consistent with the possibility that the amphibole-derived fluid beneath the volcanic front contains a greater amount of Sr and Nd than the phlogopite-derived fluid beneath the backarc-side volcanic chain. However, the widespread presence of OIB mantle sources within the mantle wedge is not generally supported by certain trace element characteristics of subduction zone lavas, such as Zr/Nb ratios which are similar to N-MORB and tend to be greater than those in OIB. Additional processes may be required to resolve these observations and models.

Finally, we point out that the sedimentary component of the subducted oceanic crust need not play an important role in controlling the *Sr-Nd-isotopic characteristics* of subduction zone magmas. In contrast, sediments are necessary to explain ^{10}Be and probably also the Pb-isotopic compositions (e.g., Tera et al., 1985; Morris and Tera, 1989; Morris et al., 1990). The fate of sediment and its role in subduction zone magmatism remains in question, particularly in view of recent claims that the incompatible trace element budgets of subduction zone magmas are correlated with and directly attributable to the amount of sediment being subducted beneath different subduction zones (Plank and Langmuir, 1993). If this proves correct, it may also have broad implications for the effectiveness of subduction in recycling sediments into the deep mantle and for the geochemical evolution of the mantle and continental crust.

5 Pressure and Temperature of Magma Generation: Origin of Across-Arc Chemical Variation

Geochemical considerations suggest that the characteristic across-arc geochemical variation, with increasing incompatible element concentrations and more alkaline magma compositions from the volcanic front toward the backarc, may be attributed to a decrease in the degree of partial melting (Chapter 3). The constraints which experimental petrology can place on the origin of these across-arc chemical systematics are examined in this chapter.

5.1 Simple System Experiments

The system diopside-forsterite-quartz (DFQ) within the simple system CaO-MgO-SiO_2 forms one of four planes of the basalt tetrahedron (Figure 3.4) and is one of the most informative ternary systems in basalt petrology. Bowen (1928) first studied this system at atmospheric pressure, and it has since been employed as an analogue for natural basalts, particularly to understand their sequences of crystallization. At high pressures, the DFQ system equally well applies to the partial melting of mantle peridotite, since each of the major phase components in peridotite (i.e., olivine, orthopyroxene, and clinopyroxene) are represented. Comparison of the liquidus phase relationships in the DFQ system at 1 atm and 2.0 GPa (Figure 5.1: after Kushiro and Schairer, 1963; Kushiro, 1969), reveals a significant contraction of the olivine liquidus field and associated shift in the position of the cotectic boundary between olivine and orthopyroxene. The change in this cotectic boundary is of major petrogenetic significance because melt compositions are required to be more olivine rich and quartz poor at

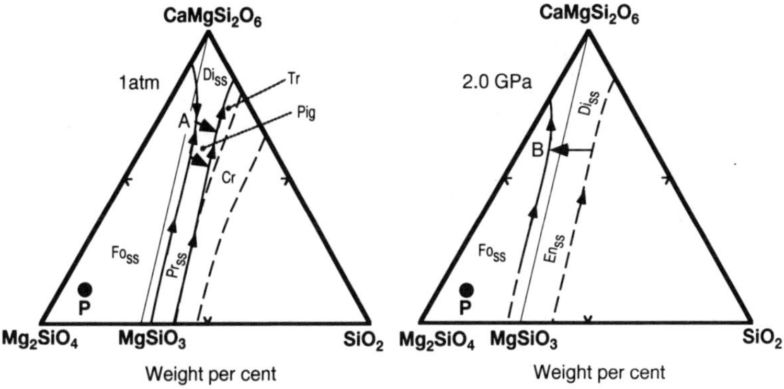

Figure 5.1 Liquidus phase diagrams for the system diopside ($CaMgSi_2O_6$)-forsterite (Mg_2SiO_4)-silica (SiO_2) at 1 atm and 2.0 GPa. Di_{ss} = diopside solid solution, Fo_{ss} = forsterite solid solution, Pr_{ss} = protoenstatite solid solution, En_{ss} = enstatite solid solution, Pig = pigeonite, Tr = tridymite, Cr = cristobalite, P = model peridotite composition, A and B are initial melt compositions of P. *(Reproduced by permission from Kushiro and Schairer, 1963; Kushiro, 1969.)*

higher pressure. The 1-atm lherzolite melt composition (A) in Figure 5.1 is SiO_2 saturated, whereas the higher-pressure melt (B) is not. It follows that higher-pressure melting of a mantle peridotite source will produce more alkaline (less tholeiitic) melt compositions.

The equilbrium partial melting behavior of a model lherzolite composition (P in Figure 5.1) can be readily demonstrated with the DFQ simple system. At an upper mantle pressure of 2.0 GPa, which is appropriate for basalt generation, the initial partial melt composition formed from a lherzolitic mantle source is located at point B. Increasing the degree of partial melting decreases the fraction of diopside (clinopyroxene) in the remaining solid until it becomes the first phase to be melted out. Further melting of the remaining harzburgitic residue drives the melt composition along the olivine-orthopyroxene boundary until olivine is the remaining solid phase, after which the melt will progress toward point P.

In addition to olivine, orthopyroxene, and clinopyroxene, fertile upper mantle peridotites also contain one or more of the aluminous phases plagioclase, spinel, and garnet. The simple system CaO-MgO-Al_2O_3-SiO_2 (CMAS) incorporates these additional phases and therefore provides a more faithful analogue for the composition, mineralogy, and melting behavior of the upper mantle. The Quaternary system

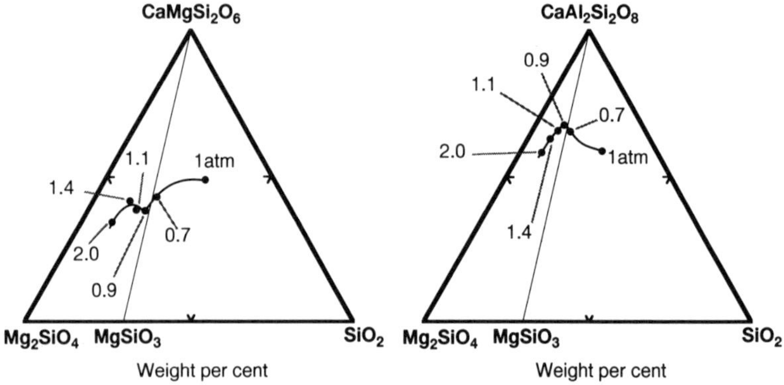

Figure 5.2 Variation of initial eutectic liquid compositions as a function of pressure (1 atm to 2 GPa) for melting of plagioclase lherzolite and spinel lherzolite mineralogies in the system CaO-MgO-Al_2O_3-SiO_2. Left side shows the projection from anorthite onto diopside-forsterite-quartz and the right side the projection from diopside onto anorthite-forsterite-quartz.

diopside-forsterite-anorthite-quartz (DFAQ) within the CMAS system is this simple system's equivalent to the subalkaline portion of the basalt tetrahedron (Figure 3.4). The effect which pressure has on partial melt compositions in the DFAQ system is readily illustrated by their projection onto the DFQ plane from anorthite and onto the FAQ plane from diopside. Figure 5.2 shows the projected experimental data of Presnall et al. (1979) and confirms that the initial liquids produced by partial melting of a mantle peridotite mineralogy change from tholeiitic to more alkaline compositions with increasing pressure. A feature of these particular projections is the discontinuity in the compositional trend that occurs at 0.9 GPa. This pressure corresponds to a low-temperature cusp (indentation) in the peridotite solidus at the transition from plagioclase-lherzolite to spinel-lherzolite mineralogy.

It might be concluded, based on the melting behavior of the simple-system peridotite analogues, that increasing pressure of melting can explain the across-arc variation in magma type, from tholeiitic at the trench-side volcanic chain to more alkaline compositions at the backarc-side volcanoes. However, simply increasing the pressure of melting will not cause an accompanying increase in incompatible element concentrations, because equivalent degrees of melting irrespective of pressure will produce melts with essentially comparable highly incompatible element concentrations.

5.2 Partial Melting of Complex/Natural Peridotite Compositions

Simple systems have contributed greatly to our understanding of basalt magma genesis, but because they do not produce real magma compositions an accurate assessment of the conditions under which a specific melt composition may be formed is not possible. To achieve this it has been necessary to conduct melting experiments on complex natural compositions and thereby include the additional effects of FeO, TiO_2, and the minor but important alkali elements Na_2O and K_2O.

Melt compositions generated in natural peridotite melting experiments were first determined with an electron microprobe at the *Geophysical Laboratory*, Carnegie Institute of Washington, by Kushiro and colleagues (1972), Mysen and Boettcher (1975a, b), and others. However, in quenching these melts to a glass phase, by rapidly cooling from the experimental conditions for usually less than 10 s, solid phases

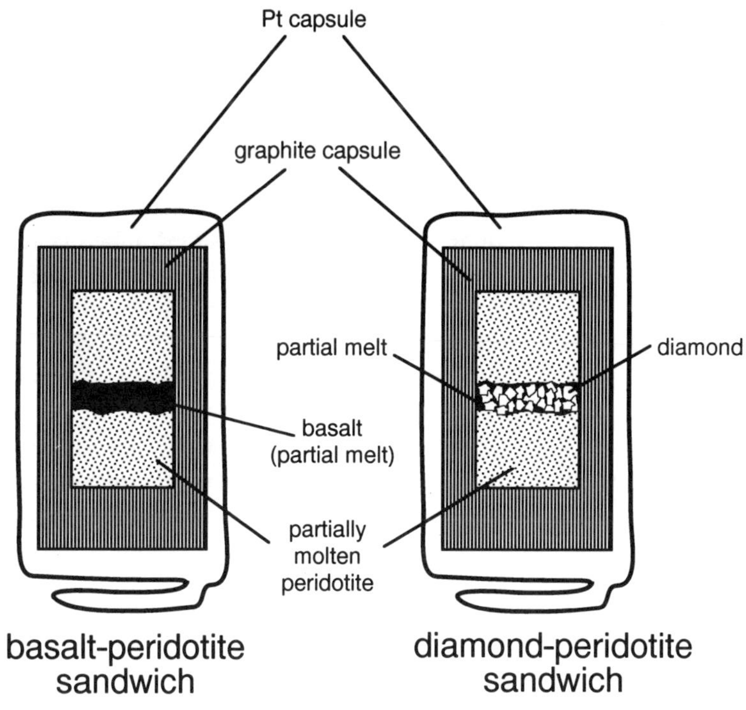

Figure 5.3 Schematic designs of basalt-peridotite and diamond-peridotite sandwich experiments.

developed overgrowths which significantly modified the partial melt compositions. This problem was partly overcome by Jaques and Green (1980), who employed mass-balance constraints to estimate interstitial melt compositions, based on modal-phase proportions and solid-phase compositions. The success of this methodology was limited by uncertainties in estimates of the modal-phase proportions.

Developments in experimental methods, particularly the peridotite-basalt "sandwich" technique (Takahashi and Kushiro, 1983; Fujii and Scarfe, 1985; Falloon and Green, 1987, 1988; Falloon et al., 1988), have subsequently facilitated the successful quenching and determination of peridotite partial melt compositions at high pressures. The schematic design of a typical "sandwich" experiment is shown in Figure 5.3. It has been demonstrated that basalt, and even andesite, layers embedded between peridotite layers will be homogeneous and have largely equilibrated with the residual peridotite phases within 24 hr (Takahashi and Kushiro, 1983).

Peridotite melting studies using the sandwich technique have produced a consistent set of data from different peridotite compositions. The consistent and systematic behavior of melt compositions, when *projected from diopside onto the plagioclase-olivine-quartz plane* of the basalt tetrahedron, enabled Falloon and Green (1988) to construct a pressure-sensitive melting grid, which is capable of establishing the depth of magma separation for primary magma compositions within at least the silica-saturated portion of the basalt tetrahedron. However, this grid has not been calibrated using data for degrees of melting smaller than ~10%, where elevated and variable incompatible minor element concentrations, arising from variations in degree of melting and starting compositions (basalt layer and peridotite sandwich) may cause significant changes to phase relationships.

A new technique, which employs aggregates of diamond embedded between peridotite layers (Figure 5.3), has the potential to overcome the small-degree melt problem (Baker et al., 1992; Johnson and Kushiro, 1992; Kushiro and Hirose, 1992; Hirose and Kushiro, 1993). In these experiments, small-degree partial melts are able to be extracted from the peridotite layers into the pore space between diamond aggregates and therein analyzed with an electron microprobe. The compositions of partial melts produced by this technique (Figure 5.4) promise improved estimates of primary magma compositions formed as a function of varying upper mantle *P-T* conditions. The results of

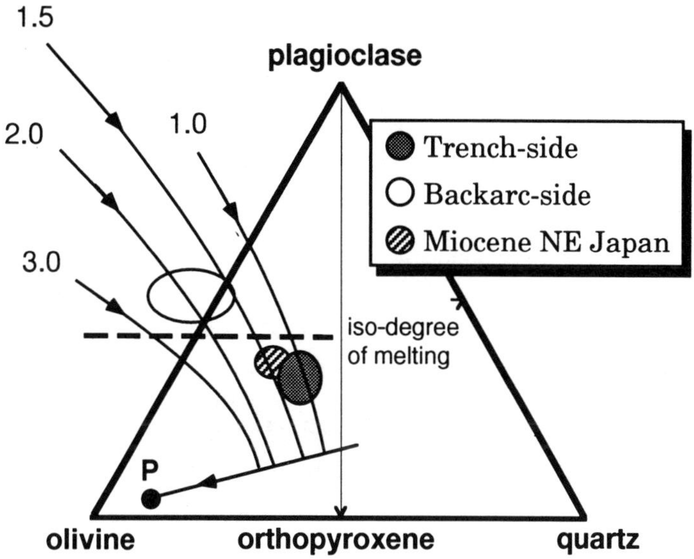

Figure 5.4 Isobaric (in GPa) melt composition trends occurring with increasing degrees of melting determined in diamond-peridotite sandwich experiments by Hirose and Kushiro (1993). Projection is from diopside onto plagioclase-olivine-quartz. *(Reproduced by permission from Walker et al., 1979.)* NE Japan arc primary magma composition estimates are shown for comparison. The dashed line indicates the approximate position of an equivalent degree of melting (~15–20%) for the various partial melting trajectories.

Hirose and Kushiro (1993) are consistent with the interpretation based on the basalt-peridotite sandwich technique and simple system experiments that more alkaline magmas will form at higher pressure given an identical degree of partial melting.

It is possible to demonstrate a difference in the depth of magma generation/separation for examples of trench-side and backarc-side primary magmas based on the melting systematics of natural peridotite compositions. The normative composition projections of primary magma estimates for the NE Japan and Sangihe arcs in Figure 5.4 provide evidence for deeper and smaller degrees of melting for the backarc-side than for trench-side magmas in both cases. Figure 5.4 can also be used to demonstrate that the lack of across-arc variation in incompatible elements observed for lavas in the Miocene NE Japan arc may be the result of similar depths of melt separation and equivalent degrees of partial melting across the arc at that time.

Both theoretical and geochemical studies indicate that very small melt fractions (<<2%) may be segregated from mantle peridotite and have sourced the suggestion that mantle melting may proceed via a fractional melting process rather than equilibrium batch melting (e.g., McKenzie, 1984; Klein and Langmuir, 1987; McKenzie and Bickle, 1988; Riley and Kohlstedt, 1991; Kinzler and Grove, 1992; Sobolev and Shimizu, 1993). This brings into question the validity of applying the results of equilibrium batch melting experiments on peridotites to estimate the pressure-temperature conditions of subduction zone primary magmas. However, Hirose and Kawamura (1994) have shown that the integrated composition of melt increments produced by fractional melting are almost identical to those produced by equivalent extents of equilibrium batch melting. Nonetheless, an important distinction does occur between these two melting mechanisms, since the temperature required for producing these equivalent magma compositions will be ~50°C higher in the case of fractional melting than that estimated based on batch melting experiments (Hirose and Kawamura, in press).

5.3 Melting-Phase Relationships of Natural Basalt

5.3.1 Multiple saturation points

A partial melt of peridotite formed at given *P-T* conditions will, at those conditions, be saturated with the residual solid peridotite phases. Accordingly, the phase relations of a primary basalt composition at high pressure can be used to evaluate the mantle conditions under which it formed, for at those conditions it will crystallize the residual peridotite phases at its liquidus. This methodology has been employed to avoid the uncertainties in *P-T* estimates arising from simply comparing magma compositions with those produced in peridotite melting studies. It remains the only viable method of accounting for the significant effects which volatiles such as H_2O may have on subduction zone magma genesis, in the absence of reliable experimental data for the partial melting of volatile-bearing peridotite compositions. To employ the method with success, we must constrain carefully the composition of the primary magma starting material, because melting experiments on differentiated and/or mixed rocks will at best provide erroneous *P-T* estimates but may result in incorrect conclusions regarding the nature of the source material and its residual mineralogy.

An example of a high-pressure-melting study on an alkali basalt composition from Skye, which is notable for its large number of experimental data points, is shown in Figure 5.5 after R. N. Thompson (1974). Olivine is the liquidus phase at pressures less than 1.6 GPa,

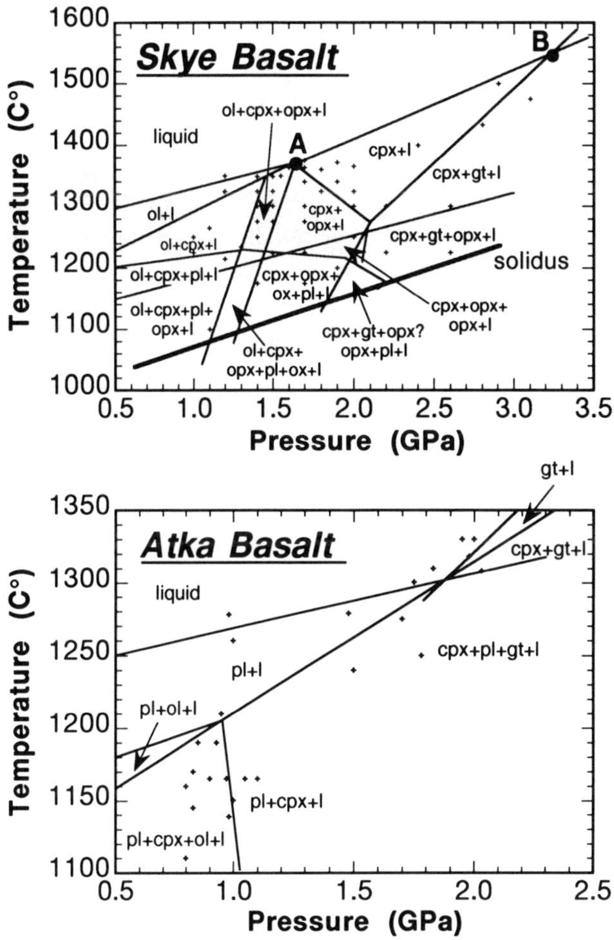

Figure 5.5 High-pressure melting phase relations for Skye and Atka basalts showing individual experiment points as small crosses. ol = olivine, cpx = clinopyroxene, opx = orthopyroxene, l = liquid, gt = garnet, ox = Fe-Ti oxide. A and B are multiple-saturation points with lherzolitic and eclogitic mineral assemblages. *(Reproduced by permission from Thompson, 1974, and Baker and Eggler, 1983.)*

clinopyroxene between 1.7 and ~3.2 GPa, and garnet at higher pressures. These relationships define two multiphase saturation points at the liquidus: the lower-pressure point has a peridotite mineralogy (olivine + orthopyroxene + clinopyroxene), and the higher-pressure point has an eclogite mineralogy (garnet + clinopyroxene). It follows that this alkali basalt composition has derived from either a lherzolite or eclogite residual mineralogy. However, because the upper mantle at such pressures is composed essentially of peridotite, it is safe to conclude, particularly given additional supporting geochemical criteria, that this basalt magma segregated from an upper mantle lherzolite source at ~1.7 GPa and 1370°C.

Some basalts studied experimentally at high pressures are not saturated with upper mantle minerals, usually because the starting compositions have not been suitable primary melt compositions. An example is shown in Figure 5.5, after Baker and Eggler (1983), who conducted melting experiments on an Aleutian (A + Ka) basalt with an FeO^*/MgO ratio of ~2 and a high content of Al_2O_3 (~20 wt%). This basalt does not crystallize olivine at its liquidus at any pressure, so this composition is unlikely to represent a primary magma produced in the upper mantle. Additional chemical characteristics, such as low-compatible-element (e.g., Ni) concentrations, support the interpretation that this basalt is derived by fractionation of olivine and accumulation of plagioclase from a more primitive basaltic magma.

No less than nine natural and synthetic subduction zone basalts have been demonstrated to be multiply saturated with essential peridotite minerals. The *P-T* conditions of their multiple saturation points under dry conditions are shown in Figure 5.6 and are compared therein to an approximate position for the solidus of mantle peridotite. This shows that multiple-saturation points comprising lherzolite assemblages plot nearer the solidus curve than harzburgite assemblages, which is consistent with the expected melting behavior of simple and natural peridotite compositions, as progressive melting results in the preferential melting out of clinopyroxene to leave harzburgite residues at higher temperatures and degrees of melting (Mysen and Kushiro, 1977).

The use of multiple-saturation points to assess the *P-T* conditions of magma generation requires that a primary magma be equilibrated at a precise *P-T* condition in the upper mantle. However, melting is a dynamic process involving melt flow relative to the melting solid material, driven by buoyancy difference and/or deviatoric stress in the

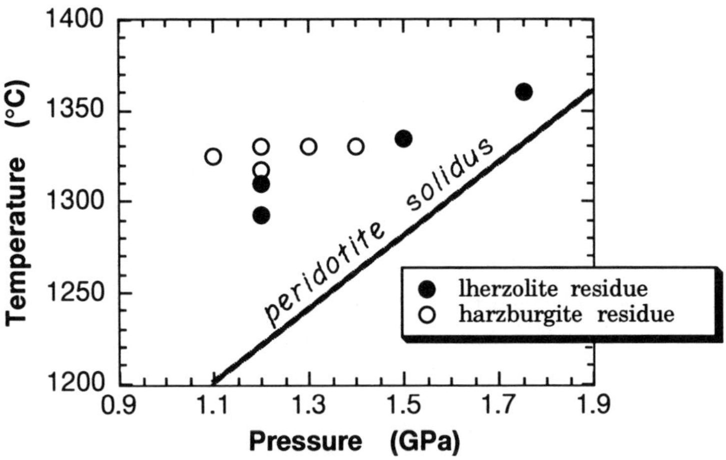

Figure 5.6 *P-T* conditions of multiple-saturation points obtained for subduction zone primary magma composition estimates. Primary magmas with harzburgitic (olivine and orthopyroxene) residues (open circles) plot at higher temperatures than those with lherzolitic (olivine, orthopyroxene, and clinopyroxene) residues (closed circles). This is consistent with higher degrees of partial melting for the former magmas. *(Data reproduced by permission from Draper and Johnston, 1992; Kushiro, 1987; Takahashi, 1980; Tatsumi et al., 1983; Tatsumi et al., 1994b.)*

mantle, to the latter of which melt is able to respond more rapidly than the solid mantle. Equilibration between solid and melt thus depends on the relative velocities of melt and solid and the rate of chemical reaction between them. Partial-melting experiments on peridotites demonstrate that equilibration between melt and mantle phases can be achieved within 24 hr (Takahashi and Kushiro, 1983; Hirose and Kushiro, 1993). Most estimates of melt velocities during melting are from centimeters to meters per year, so it seems likely that melt will maintain equilibrium with the mantle through which it flows, even on a grain-to-grain scale. Consequently, a multiple-saturation point can be considered with some confidence to reflect the *P-T* conditions where melt ceases porous flow and changes to a regime dominated by flow through propagating cracks or conduits, effectively isolating the melt from its mantle source.

5.3.2 Effect of volatiles

Volatiles such as H_2O and CO_2 are soluble in magmas at mantle pressures. They are known to cause significant changes to the chemical structure of silicate melts and, as a consequence, to melt-solid phase relations. It is essential therefore to consider the effects which these volatiles may have on the generation of subduction zone magmas, particularly given evidence for the role which H_2O plays in magma generation.

The solubility of H_2O in silicate melts has been determined from the H_2O content of glasses formed under H_2O-oversaturated conditions (e.g., Hamilton et al., 1964; Oxtoby and Hamilton, 1978), the occurrence of H_2O-saturation based on the phase relations of a magma with variable amounts of added H_2O (Yoder et al., 1957; Burnham and Jahns, 1962; Tatsumi, 1981), and the occurrence of H_2O oversaturation based on the presence of bubbles in quench glasses formed by melt decompression (Sakuyama and Kushiro, 1979). These experimental studies have shown that the temperature and compositional dependence of the solubility of H_2O is not large, but that there is a strong dependence on pressure at less than 2.0 GPa (Figure 5.7).

The solubility of CO_2 in a variety of compositions has been estimated using chemographic methods (Eggler, 1973), direct measurement of CO_2 in quench glasses by chromatographic techniques (Moore et al., 1970; Holloway and Lewis, 1974), and by beta-track mapping (Mysen et al., 1976). These experimental data indicate a strong contrast between the solubility behavior of CO_2 and H_2O, with CO_2 being sensitive to both bulk composition and temperature. Moreover the amount of CO_2 able to be dissolved is much smaller than H_2O at an equivalent pressure and temperature. The solubility of CO_2 at the liquidus temperature for two compositions is shown in Figure 5.7 as a function of pressure after Spera and Bergman (1980). The comparatively low solubility (<2 wt%) of CO_2 in typical basalt compositions at pressures less than ~2.5 GPa suggests it may have only a limited role in subduction zone magmatism.

Experimental data on the phase relations of volatile-bearing silicate systems relevant to the partial melting of upper mantle peridotites have led to the suggestion that both H_2O and CO_2 are important in controlling partial-melt compositions at high pressures. The shift in composition of initial melts formed in two simple systems in the presence of volatiles is shown in Figure 5.8. In the DFQ system at 2.0 GPa, a volatile-poor liquid in equilibrium with lherzolite (point B in Figure 5.8) is not saturated with a quartz component but is at H_2O-saturated conditions (point C). Generation of a more-silica-rich liquid

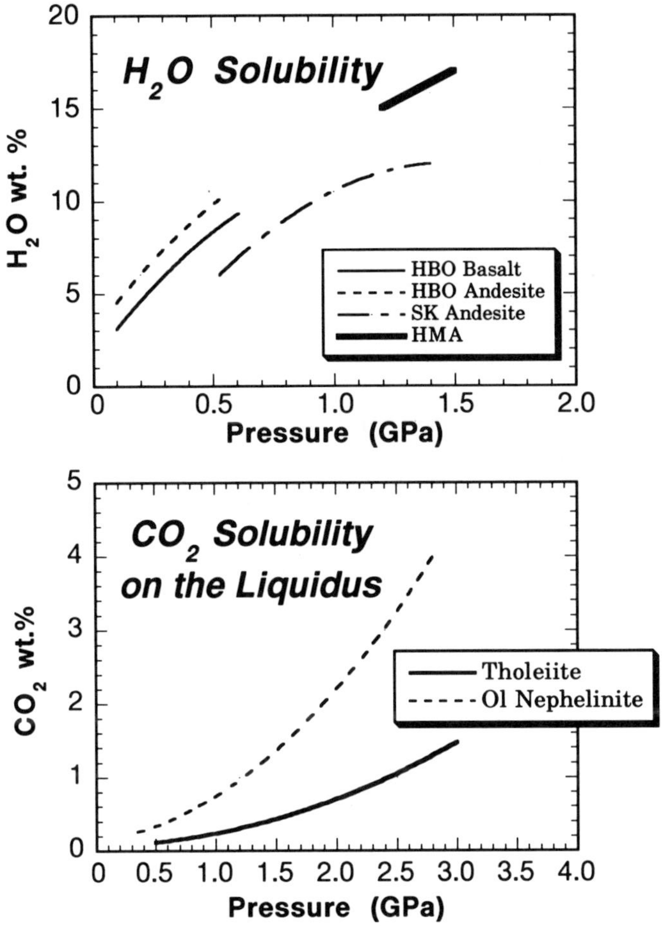

Figure 5.7 Solubility of CO_2 and H_2O in various silicate melt compositions. (*Data reproduced by permission from Hamilton et al., 1964 [for HBO basalt and andesites]. Sakuyama and Kushiro, 1979 [for SK andesite]. Tatsumi, 1981 [for HMA (high-Mg andesite)]. Spera and Bergman, 1980 [for tholeiite and olivine nephelinite].*)

in the presence of H_2O as opposed to CO_2 is also suggested in the system nepheline-forsterite-quartz (Figure 5.8), and equivalent behavior has been demonstrated in natural peridotite compositions (e.g., Mysen and Boettcher, 1975a,b). These relationships suggest that a variety of magma compositions, from quartz to nepheline-normative, may form at the same pressure from a peridotite source, simply due to the presence of different volatile components.

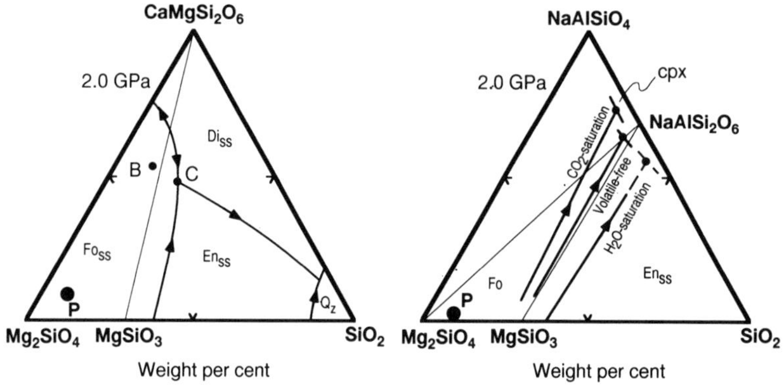

Figure 5.8 Liquidus phase diagrams for the system diopside-forsterite-silica and nepheline-forsterite-silica in the presence of H_2O or CO_2 at 2.0 GPa. P = model peridotite composition, Di_{ss} = diopside solid solution, Fo_{ss} = forsterite solid solution, En_{ss} = enstatite solid solution, cpx = clinopyroxene, Qz = quartz. *(Reproduced by permission from Kushiro, 1969.)*

Dissolved volatiles also have a profound effect on the solidus temperature of silicate systems (Kushiro et al., 1968; Green, 1973; Millhollen et al., 1974; Eggler, 1973, 1974). The solidi for a peridotite composition in the presence and absence of different volatiles are drawn schematically in Figure 5.9. The solidus temperature in the volatile-free system increases with elevated pressures, whereas under H_2O-saturated conditions it is rather constant at ~1000°C for pressures between 1 and 4 GPa, and is between 200 and 600°C lower than the volatile-free solidus. Under H_2O-undersaturated conditions, crystallization of pargasitic amphibole may buffer H_2O in a peridotite system and govern its melting behavior at pressures less than ~2.8 GPa (Figure 5.9). The solidus temperature of a peridotite in the presence of CO_2 vapor increases with pressure up to ~2 GPa but decreases relative to the anhydrous solidus, as the solubility of CO_2 in the silicate melt increases with pressure (Figure 5.7). The CO_2-saturated peridotite solidus temperature decreases sharply between 2.0 and 2.5 GPa and then increases again at higher pressures due to the stablization of solid carbonate phases (Figure 5.7). The initial lowering of the solidus temperature accompanies a change to more CaO-rich melt compositions.

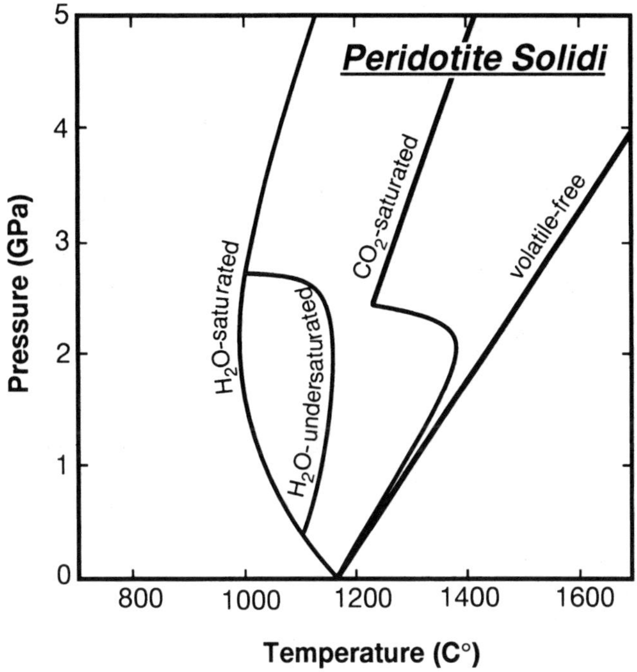

Figure 5.9 Solidus positions for peridotite in the presence and absence of H_2O and CO_2.

Melting-phase relations for natural and synthetic basalt compositions (Tatsumi, 1982; Tatsumi et al., 1983) show that no significant changes in phase relations in the presence of H_2O but do show a shift in the multiple-saturation point to higher pressure and lower temperature (Figure 5.10). This reflects a partial cancellation of the effect of H_2O by that of pressure, and indicates that a given quantity of H_2O can be balanced by a pressure increase, such that the interpretation of a melt composition deriving from a peridotite based on volatile-absent considerations may not change apart from a higher H_2O content requiring a higher pressure of melting. The effect of CO_2 is opposite that of H_2O (Figure 5.10), with the multiple-saturation point shifting to lower pressure and temperature due to expansion of the orthopyroxene liquidus field relative to that of olivine. The preferential stabilization of orthopyroxene is illustrated in melting experiments on natural alkali basalts (e.g., Iwamori, 1991), wherein alkali basalt

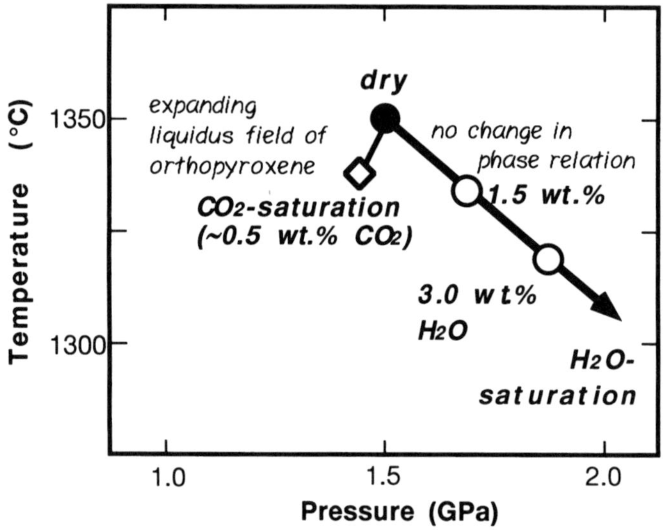

Figure 5.10 The shift in multiple-saturation point of a subduction zone primary magma composition occurring in the presence and absence of H_2O and CO_2.

compositions saturated with olivine and clinopyroxene in the absence of CO_2 have been saturated at their liquidi with orthopyroxene under nearly-CO_2-saturated conditions. We suggest these effects of CO_2 may be limited in subduction zones, due to the relatively low solubility of CO_2 in basaltic melts at the pressures of melt equilibration in the mantle wedge (Figure 5.7).

5.4 *P-T* Conditions of Magma Separation in Subduction Zones

5.4.1 Basalt magmas

Both simple system and peridotite melting experiments indicate that more tholeiitic and generally lower-K magmas are produced at shallower levels and by greater degrees of partial melting (see Figure 5.4). This is supported by experimental results including the multiple-saturation points of specific basalt compositions within a single volcanic arc (Figure 5.11). Primary melt estimates for trench-side basalts in the Quaternary NE Japan arc are multiply saturated with harzburgitic minerals at ~1.1 GPa and 1320°C, whereas backarc-side magmas are

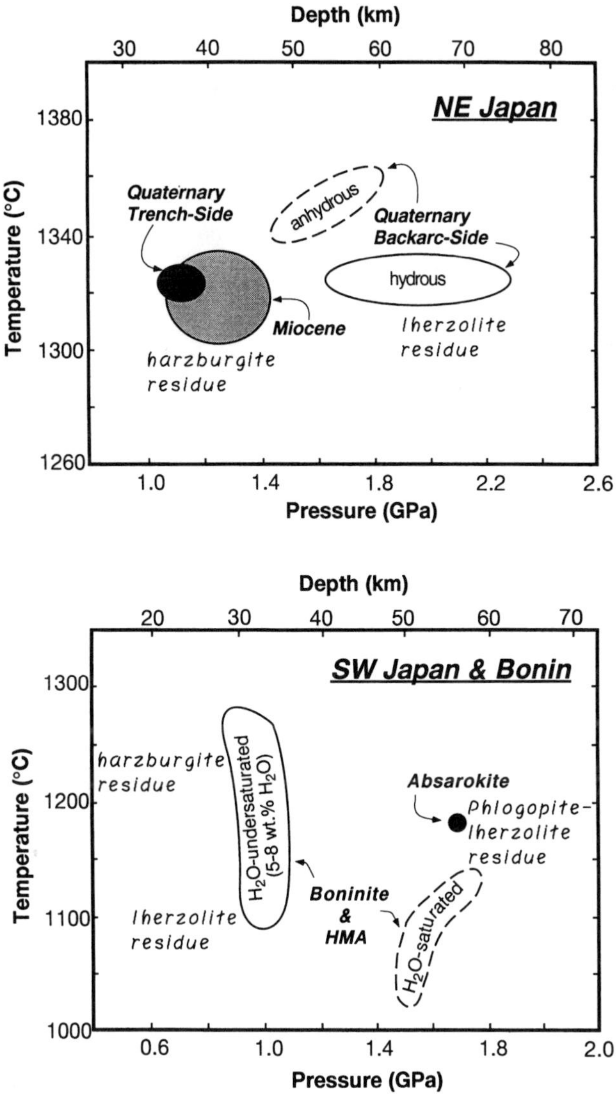

Figure 5.11 *P-T* conditions of magma separation deduced for the NE Japan, SW Japan, and Bonin arcs based on experimentally determined multiple-saturation points of basaltic, high-Mg andesite, and absarokite primary magma estimates.

saturated with lherzolitic minerals at higher pressures (1.4–1.8 GPa) under anhydrous conditions. These conditions apply to the final equilibration between a primary magma and its residual mantle materials and are consistent with estimates based on peridotite partial-melting experiments (e.g., Takahashi and Kushiro, 1983; Hirose and Kushiro, 1993). Assuming that 1–3 wt% H_2O is contained in backarc-side primary magmas, as argued in Chapter 3, the conditions of magma generation are estimated to be 1.5–2.2 GPa and ~1320°C (Figures 5.10 and 5.11). These estimates imply that the observed across-arc variation in incompatible element concentrations and magma type accompanies deeper magma separation and smaller degrees of partial melting toward the backarc.

It is notable that several recent models for subduction zone magmatism have proposed temperatures in the mantle wedge which are considerably lower than the estimates derived from liquidus experiments on primary magma compositions. For example, Davies and Bickle (1991) suggested 1250°C to be the maximum temperature attained in the mantle wedge. However, if observed primary magma compositions are to be produced at such low temperatures, the presence of considerable amounts of H_2O will be needed, with possibly up to ~10 wt% H_2O dissolved in the melt (see Figure 5.10). Such hydrous magmas will become saturated with H_2O during ascent at pressures of the range 0.5–1.0 GPa and will undergo degassing and rapid crystallization at these pressures (see Figure 5.7). The eruption of magnesian, aphyric (crystal-free) basalts (e.g., Eggins, 1993) argues against this possibility, as does the lack of direct evidence for very high H_2O contents, such as the relative scarcity of amphibole phenocrysts in subduction zone magmas, even in evolved magma compositions.

The mechanism via which melt transport takes place in the region of melt production, prior to melt separation from the mantle and migration to the surface, is probably dominated by porous flow along solid grain boundaries (McKenzie, 1984). In contrast, magma separation from the mantle occurs by buoyant ascent through cracks and fractures (Anderson and Grew, 1977; Sleep, 1988; Spence and Turcotte, 1990; Takada, 1990; Lister and Kerr, 1991, Furukawa, 1993). The presence of dikes in mantle peridotites testifies to the importance of melt-filled crack propagation under relatively shallow lithospheric mantle conditions (Shaw, 1980; Nicolas and Jackson, 1982; Nicolas,

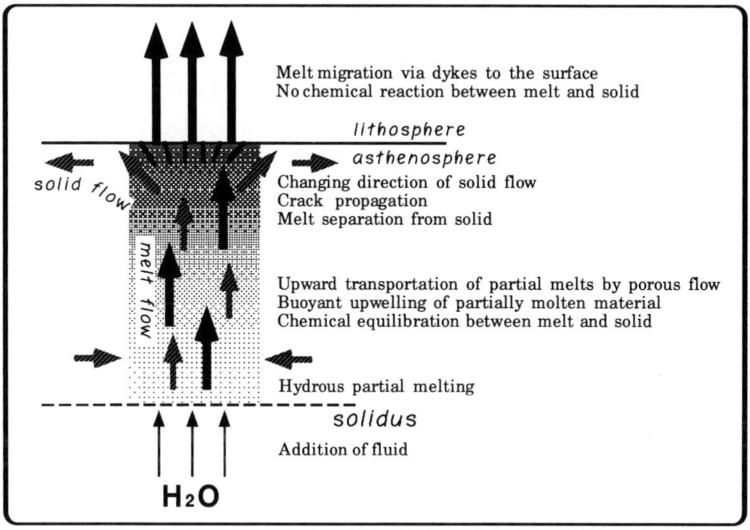

Figure 5.12 A schematic diagram illustrating the process of melt migration and separation process within the mantle wedge.

1986). We suggest that the chemical equilibration or multiple-saturation point of a magma reflects the *P-T* conditions of magma separation from the mantle or, equivalently, the conditions where the change from porous flow to crack dominated melt flow takes place. This change is likely to isolate effectively a magma from further chemical equilibration with the mantle, and relatively little potential may exist for chemical interaction between the magma and the surrounding/overlying mantle materials after that point (Figure 5.12). Crack creation and growth may be governed by local differential stress, such that when rising partially molten mantle reaches a more rigid (higher viscosity) overlying region, ascent is curtailed and the solid flow may be forced horizontally, facilitating crack formation and separation of the buoyant magma (Figure 5.12). The viscosity of mantle materials is essentially controlled by their temperature, and the location of a sufficiently large increase in the viscosity may therefore take place near a particular isotherm at or near the asthenosphere/lithosphere boundary. We speculate that the decrease in degree of partial melting toward the backarc may result from increasing depth to the asthenosphere/lithosphere boundary in that direction. This interpretation is harmonious with seismological observations (see Figure 2.2).

The high-pressure melting-phase relations of Mg-rich lavas from the Miocene NE Japan arc indicate no systematic difference in depth of magma separation in the across-arc direction at that time (Yamashita and Fujii, 1992; Tatsumi et al., 1994b), being ~1.4 GPa for both the trench- and backarc-side volcanoes. This contrasts with ~1.0 GPa for the trench side and >2.0 GPa for the backarc side in the Quaternary NE Japan arc (Figure 5.11). These observations, together with the outlined concept of magma separation at the lithosphere/asthenosphere boundary, have led Tatsumi et al. (1994b) and Yamashita and Tatsumi (1994) to the conclusion that the unusual across-arc variation in K_2O results from thinning of the lithosphere beneath the backarc side at the time of opening of the Japan Sea backarc basin during the late Oligocene to early Miocene. This explanation may also apply to the reverse pattern of K_2O variation in the Banks Islands, Vanuatu, due to rifting in the Vot Tande Trough and active backarc rifting in the northwestern sector of the North Fiji Basin.

5.4.2 Other subduction zone magmas

Experimental studies in simple systems (e.g., O'Hara, 1965; Kushiro, 1972, 1974) demonstrate that partial melts of hydrous peridotites are enriched in SiO_2 compared with their anhydrous counterparts. Mysen and Boettcher (1975a) first showed that such silica-rich melts have broadly andesitic compositions, and more recent studies on high-Mg andesites have concluded that partial melting of hydrous peridotites *does* take place in nature to produce high-Mg andesite magmas, such as the boninites in the Bonin Islands (Kuroda et al., 1978), sanukitoids in SW Japan (Tatsumi and Ishizaka, 1981, 1982a,b), and clinoenstatite-bearing rocks from Cape Vogel, Papua, New Guinea (Dallwitz et al., 1966).

Melting-phase relations of high-Mg andesites (Kushiro and Sato, 1978; Tatsumi, 1981, 1982; Umino and Kushiro, 1989; Van Der Laan et al., 1989) demonstrate that high-Mg andesite magmas are multiply saturated with harzburgitic or lherzolitic mineral assemblages under both H_2O-oversaturated and -undersaturated conditions. It is possible to produce high-Mg andesite magmas from anhydrous peridotites at pressures close to 1 atm; however, given an upper mantle origin for such magmas, pressures around 0.5 GPa or more are more reasonable estimates. The generation of high-Mg andesite magmas under H_2O-saturated conditions is also unlikely, since this would result in large-scale crystallization upon ascent and cause complete solidification at depth, because the liquidus temperatures of H_2O-saturated andesites are less than their solidus temperatures at very low pressures of eruption. It can be concluded that high-Mg andesites most probably form at relatively shallow depths in the uppermost mantle under H_2O-undersaturated conditions (Figure 5.11). Recent Sr-Nd-Pb isotope studies on high-Mg andesites in SW Japan have demonstrated a large contribution from subducted sediments in their formation. It is most likely that the mass introduction of such sediment components has been effected by partial melts of the subducted oceanic crust rather than sub-derived fluids.

A very-high-K, absarokitic basalt from the SW Japan arc has also been studied experimentally by Tatsumi and Koyaguchi (1989) and found to have a multiple-saturation point with phlogopite, olivine, orthopyroxene, and clinopyroxene at ~1.7 GPa and 1170°C (Figure 5.11). Melting experiments on another alkali basalt composition from the SW Japan arc, in the presence of CO_2 and H_2O (Iwamori, 1991), provide estimated *P-T* conditions for magma separation of ~1.8 GPa and 1320°C. However, it remains unclear as to whether these particular magmas in SW Japan can be directly related to subduction processes.

6 Composition of Subduction Zone Fluid Phases: An Experimental Assessment of Geochemical Characteristics

6.1 Metasomatic Agents

It was shown in Chapter 3 that primitive subduction zone magmas are relatively enriched in incompatible trace elements which have larger ionic radius/charge ratios or lower ionic potential. Moreover, none of the possible source components present within subduction zones (i.e., sediments, original mantle wedge, or altered oceanic crust) match these characteristics sufficiently closely to be considered the source of their origin. Thus, the difference in chemistry of the mantle wedge source of subduction zone magmas compared with other parts of the upper mantle can be considered to reflect the selective metasomatic addition of particular elements from the subducting oceanic lithosphere. Identification of such subduction-derived chemical components has been a prime target of petrological and geochemical studies of subduction zone magmas.

What is the nature of the metasomatic agent(s) which enriches the source region of subduction zone magmas? Since first proposed by Nicholls and Ringwood (1973), many petrologists have favored a mechanism involving melting of the subducting oceanic crust and the subsequent reaction of these melts with the overlying mantle wedge. Sekine and Wyllie (1982) suggested this to be "an attractive and versatile hypothesis, consistent with recent geochemical evidence in broad outline." An alternative is that fluids released from dehydration reactions in the subducted lithosphere may metasomatise and cause partial melting of the mantle wedge (e.g., Coats, 1962). A significant majority of researchers currently believe that the subducted oceanic crust does not melt under normal circumstances, based on several lines of reasoning. (1) The dipping seismic plane is usually continuous to depths of ~200 km—that is, from subforearc through, and beyond, the subvolcanic arc (Chapter 2). (2) The constant depths to the surface of the subducted oceanic lithosphere beneath the chains of arc volcanoes indicate that the occurrence of pressure-dependent reactions, not temperature-dependent melting processes, are likely to characterize melting of the subducted oceanic crust (Chapter 2). (3) Certain subduction zone magma compositions, which have been claimed to be partial melts of the oceanic crust (e.g., high-alumina basalt magmas; Johnston, 1986; Brophy and Marsh, 1986), have been demonstrated to be the differentiation products of more primitive melts generated in the mantle wedge or the result of plagioclase phenocryst accumulation in magmas of basaltic andesite to dacite compositions (e.g., Crawford et al., 1987). These observations compel us to propose that the subducted oceanic lithosphere does not melt but rather undergoes subsolidus dehydration reactions which release fluid phases that play an essential role in chemical enrichment and partial melting of the mantle wedge. However, we do need to qualify this position for subduction zones with unusually high-temperature conditions may result in partial melting of the oceanic crust—for example, where subduction of young and hot oceanic lithosphere takes place (e.g., Drummond and Defant, 1990; Peacock, 1990) and which may have existed more widely during early Earth history.

The principal species in a fluid phase present in the upper mantle in the system C-O-H can be calculated from fluid equations of state, such as the Redlich–Kwong equation (e.g., Holloway, 1981; Eggler, 1987). The fluid species present are sensitive to oxygen fugacity (f_{O_2}), such that under relatively oxidizing conditions in the mantle (near or above the fayalite-magnetite-quartz (FMQ) and wustite-magnetite (WM) oxygen buffers) fluids are H_2O-CO_2 dominated, whereas under more reducing mantle conditions (approaching the iron-wustite (IW) buffer) they are CH_4-H_2O dominated. Large variations in the f_{O_2} of the upper mantle have been proposed (e.g., Arculus, 1985) and are evidenced by the occurrence of Fe-Ni metallic alloys enclosed by diamond in kimberlites, which require reducing conditions near the IW buffer (Bulanova et al., 1979), and by ilmenite-spinel intergrowths, which also occur as kimberlite-hosted inclusions but indicate oxidizing conditions near the FMQ buffer (Haggerty and Tompkins, 1983). Application of new oxygen barometers to olivine-orthopyroxene-spinel assemblages in mantle peridotites also indicates significant variation in the oxidation state of the mantle and derivative magmas (Haggerty, 1990; Ballhaus et al., 1991; Ballhaus, 1993). However, several lines of evidence indicate that subduction zone magmas are characterized by relatively high oxygen fugacities: (1) the chemistry of olivine-spinel pairs in both lavas and cumulate xenoliths in subduction zone settings indicates f_{O_2} values ranging from above FMQ to FMQ plus 3 log units, as compared with ~FMQ minus 1 log unit to FMQ for MORB (e.g., Ballhaus, 1993); and (2) ferromagnesian phenocrysts in primitive subduction zone basalts are often more magnesian than those in primitive MORB and ocean island basalts, possibly reflecting higher initial Fe^{3+}/Fe^{2+} ratios in primary magmas (Ramsay et al., 1984). It is not yet clear whether the oxidized nature of subduction zone basalts is inherited from their source region, since a variety of processes may shift a magma's f_{O_2} relative to that of its source region, including melt segregation processes during partial melting, varying Fe^{3+}/Fe^{2+} ratios during melt decompression from the source region, oxygen exchange with the lithospheric mantle or crust during magmatic ascent, and low-pressure volatile degassing. Taking into account these possibilities, it has been concluded that the f_{O_2} in the mantle wedge is near or above the FMQ buffer (Arculus, 1985; Ballhaus, 1993), requiring any C-O-H fluid phase present to be a H_2O-CO_2 mixture.

CO_2 may be present in mantle fluid phases in subduction zones; however, most evidence suggests that these fluids are dominated by H_2O, including the compelling evidence for dehydration reactions being responsible for the formation of two volcanic chains (Chapter 4).

6.2 Composition of Aqueous Fluids at Upper Mantle Conditions

What role do aqueous fluids play in transporting elements and determining geochemical fluxes within subduction zones? Under upper mantle *P-T* conditions, H_2O exists as a supercritical fluid which can dissolve additional components and elements to a certain extent. Goldschmitt(1954) showed that the ionic potential of an element—that is, the ratio of its charge and ionic radius—is a critical parameter in determining the solubility of an element in an aqueous fluid. This effectively divides elements into two groups, those which are relatively soluble in H_2O and those which are not (Figure 6.1). The large-ion lithophile elements, which have small ionic potentials, will be more readily dissolved and transported by an aqueous fluid than are high-field-strength elements, which have large ionic potentials. Elements which have extremely large ionic potentials, such as boron, may also be soluble in H_2O, because they can form anionic complexes with oxygen in an aqueous solution. The degree to which an element will dissolve into a fluid will also be a function of its compatibility in the solid phases with which the fluid coexists.

The solubility of SiO_2 in an aqueous fluid in the presence of peridotitic minerals has been estimated by Nakamura and Kushiro (1974). Figure 6.2 shows an isothermal section of the system H_2O-Mg_2SiO_4-SiO_2 at 1.5 GPa and indicates a marked variation in the solubility of SiO_2 with change in the phases which coexist with the fluid or gas phase. The H_2O fluid/gas contains up to 20 wt% SiO_2 in the presence of olivine, whereas it contains 20–40 wt% SiO_2 in the presence of enstatite. Therefore, in a simple peridotite system containing olivine and enstatite, an aqueous fluid at upper mantle conditions (strictly 1.5 GPa and 1280°C) is able to dissolve ~20 wt% SiO_2.

Crystal-fluid partitioning experiments by Schneider and Eggler (1986) further demonstrate a range in bulk partition coefficients for various chemical components between aqueous fluids and the assemblage forsterite-enstatite-phlogopite at pressures between 1.3 and 2.2 GPa (Figure 6.3). Experiments have also documented that

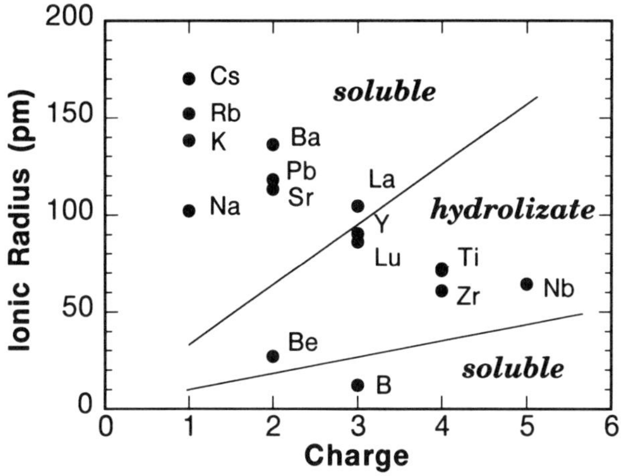

Figure 6.1 Diagram illustrating predicted soluble and hydrolizate (insoluble) element behavior in supercritical aqueous fluids based on ionic potential.

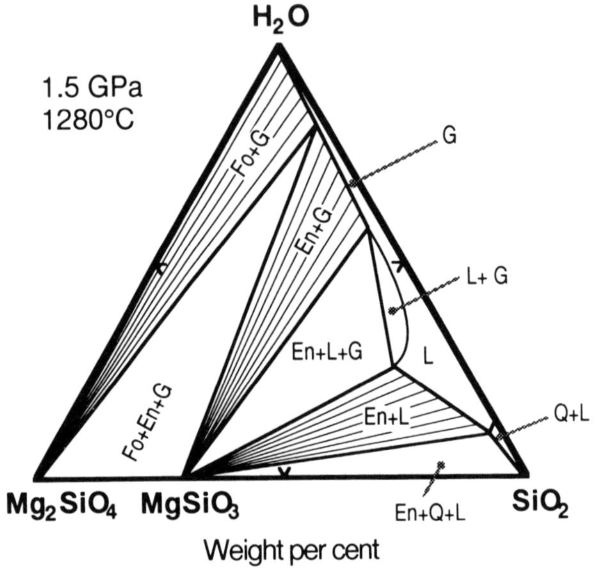

Figure 6.2 Isothermal section of Mg_2SiO_4-SiO_2-H_2O at 1.5GPa and 1280°C. Fo = forsterite, En = enstatite, L = liquid, G = gas or vapor. *(Reproduced by permission from Nakamura and Kushiro, 1974.)*

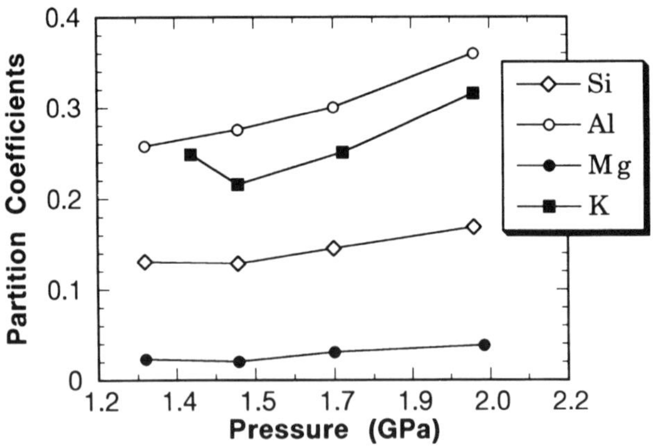

Figure 6.3 Bulk partition coefficients between aqueous fluids and the assemblage forsterite-enstatite-phlogopite. *(Reproduced by permission from Schneider and Eggler, 1984.)*

markedly higher solubilities occur in H_2O fluids than in CO_2-rich fluids. Figure 6.3 shows that K_2O and Al_2O_3 can be highly soluble in an aqueous fluid under upper mantle conditions. The higher solubility of K_2O has also been demonstrated experimentally at 3.0 GPa by Ryabchikov and Boettcher (1980). Finally, the partitioning of a range of trace elements between olivine and H_2O at high pressures was determined by Brenan and Watson (1991) and showed comparatively higher solubility of Cs, Ba, and Sr than of REE in H_2O fluids.

The comparative solubility and ease of transportation of elements in an aqueous fluid formed by decomposition of hydrous phases at mantle pressures and temperatures has been determined by comparing the composition of hydrous materials, in particular natural serpentinite and synthetic serpentine, with their experimentally dehydrated residues, which comprise olivine and orthopyroxene (Tatsumi and Nakamura, 1986; Tatsumi et al., 1986; Tatsumi and Isoyama, 1988; Kogiso et al., unpublished data). The mobility of an element can be expressed simply as the ratio of the mass of an element lost during the dehydration process to that originally present. Examples of experimental data produced in our laboratory are summarized in Figure 6.4. Because many of the elements shown may be distributed principally on grain boundaries in the serpentine and not bound in the crystal structure, the experimental results may reflect partitioning of elements between grain boundaries and an aqueous fluid rather than residual minerals and the fluid. Nonetheless the data in Figure 6.4 indicate that incompatible elements with larger ionic radii are more readily transported by a H_2O fluid compared to smaller elements of the same charge. Exceptions to this behavior occur for lead and boron, with preferential transportation of B relative to other trivalent cations and an apparently low degree of mobility of Pb compared to other divalent cations.

The fluid-mobility data derived from serpentine dehydration experiments may not strictly apply to natural reactions occurring in subduction zones, because the residual phases produced by serpentine decomposition (i.e., olivine and orthopyroxene) differ from those present following decomposition of amphibole, phlogopite, chlorite, and so on. However, most elements, with the exception of the HREE, which may be strongly partitioned into garnet during fluid-mineral reactions, are likely to be relatively incompatible with respect to most residual phases, suggesting that element mobilities inferred from the

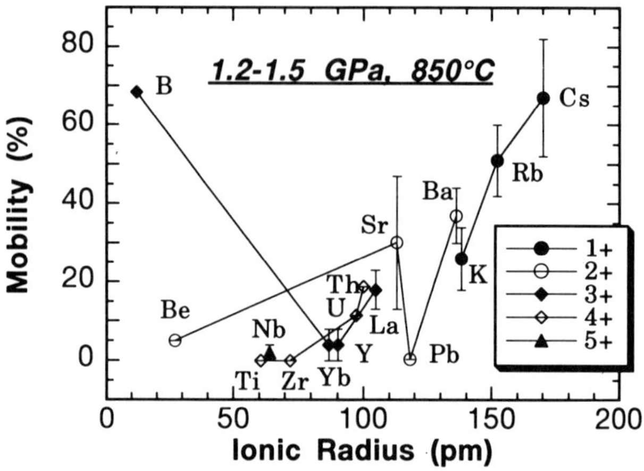

Figure 6.4 Trace element mobility as a function of ionic radius, as determined from experimental decomposition of serpentine at 1.2–1.5GPa and 850°C.

serpentine dehydration experiments could be applied at least crudely to the natural reactions.

The detection of ^{10}Be in subduction zone lavas verifies a contribution from oceanic sediments to the production of arc magmas (e.g., Brown et al., 1982; Tera et al., 1986). Be has been believed to be transported to the magma source region in the mantle wedge by partial melts of the subducted crust, based on its high ionic potential and the resultant expectation that it will be insoluble in an aqueous fluid phase (Figure 6.1). However, experimental results (Tatsumi and Isoyama, 1988) indicate that ~5% of Be is extracted along with H_2O during dehydration of serpentine (Figure 6.4), and it follows that slab melting is not necessary to account for the inclusion of ^{10}Be in subduction zone lavas.

Partitioning of elements between an aqueous fluid and a silicate melt may also contribute, at least partially, to the geochemical characteristics of arc magmas, since adding H_2O to the partially molten zone in the mantle wedge can result in H_2O oversaturation in that region and formation of a separate vapor phase in the presence of partial melt. Decompression experiments on hydrous andesite compositions, in which a H_2O vapor phase forms and is separated from the melt, demonstrate that a H_2O vapor phase may be effective in transporting Na_2O (Sakuyama and Kushiro, 1979). The mobility of some trace

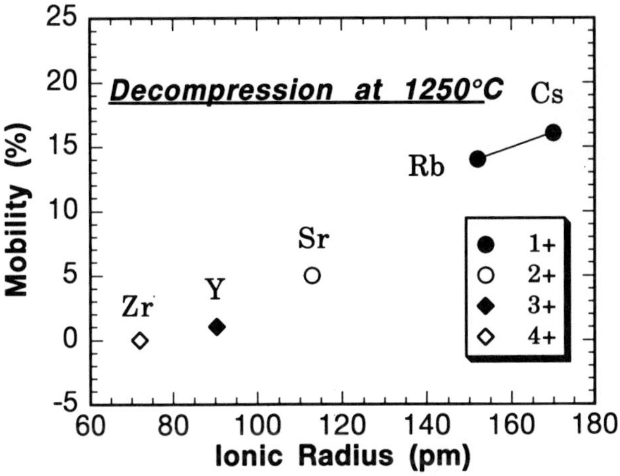

Figure 6.5 Element mobility as a function of ionic radius during volatile exsolution from a basaltic melt.

elements through volatile exsolution from a basaltic melt has been measured by the same method and has revealed preferential transport of Cs and Rb compared to Y and Zr in the fluid phase (Figure 6.5; Tatsumi and Yoshida, unpublished data). Eggler (1987) conducted high-pressure experiments which also show higher H_2O-melt partition coefficients for Rb, K, and Na than Zr, Nb, and Ti. This data is consistent with other available data for aqueous fluid-solid partitioning and element transport by a vapor phase migrating within/from a silicate melt.

Despite the need for further data on fluid-solid and fluid-melt element partitioning, the experimental results outlined indicate that the diagnostic geochemical characteristics of subduction zone magmas—that is, their relative enrichments in incompatible elements with small ionic potential—can be reasonably attributed to the preferential solubility and transport of these elements by aqueous fluids, which are derived from dehydration reactions both in the subducting oceanic crust and the mantle wedge. A possible role for volatile exsolution processes within fluid-saturated parts of the partially molten mantle wedge may also need to be considered.

7 Mantle Wedge Dynamics

Experimental phase equilibria provide a static view of basalt magma genesis in subduction zones: pressure-sensitive dehydration reactions at the base of the mantle wedge can explain the formation of two volcanic chains in volcanic arcs, the preferential solubility of elements with low ionic potential in aqueous fluids can explain the distinctive incompatible element chemistry of subduction zone magmas, and deeper and smaller degrees of melting toward the backarc can explain the across-arc variation in magma compositions. But how do these discrete phenomena integrate with the dynamic processes of melting and of mantle flow which must take place in the mantle wedge, and do dynamic processes influence the generation of magma? Moreover, do dynamic processes contribute to the resolution of the paradox of subduction zone magmatism—that is, the occurrence of melting and high heat flow in a setting where cold oceanic lithosphere sinks into and cools the mantle?

7.1 Petrological Constraints on Mantle Wedge Dynamics

7.1.1 Viscosity of the mantle wedge

Material that flows in response to an applied stress is a fluid. Mantle materials including both unmolten and partially molten peridotites are fluids, and their fluid behavior facilitates plate tectonics and plays a major role in determining the thermal and chemical evolution of the Earth. The motion of a fluid can be described by the Navier-Stokes equation in which the Reynolds number, the ratio between the inertia and viscosity terms, determines the flow behavior, and the Rayleigh number, which is inversely proportional to viscosity, controls the onset and style of thermal convection. It is clear that viscosity plays a key role in determining the dynamic behavior of fluids. The viscosity of the Earth's mantle has been constrained by studies of the dynamic response of the mantle to loading and unloading at the surface, such as during postglacial rebound. A viscosity of 4×10^{19} Pa·s has been estimated for the asthenospheric mantle by Cathles (1975); however, the mantle wedge viscosity (η) will not be constant but will vary as a function of pressure and temperature:

$$\eta = \eta_0 \exp\left(\frac{E + PV}{RT}\right)$$

where η_0 is a proportional constant, E is the activation energy, P is the pressure, V is the activation volume, R is the gas constant, and T is the absolute temperature. The pressure- and temperature-dependent behavior of the mantle viscosity is shown in Figure 7.1, based on the set of following values which realize a viscosity of 1×10^{19} Pa·s at 3.0 GPa and 1400°C, consistent with the asthenosphere value estimated by Cathles; that is, E = 420 kJ mol^{-1}, which is appropriate for a damp mantle (Karato et al., 1986), $V = 17 \times 10^{-6}$ m^3 mol^{-1} (Kirby, 1983), and

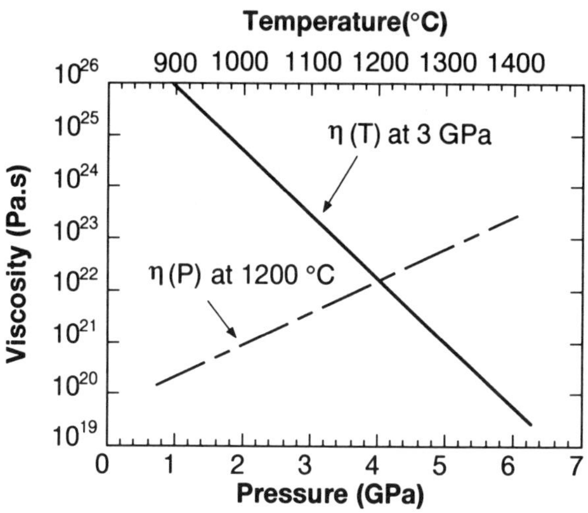

Figure 7.1 Illustration of the pressure- and temperature-dependent behavior of mantle viscosity. $\eta(P)$ = isothermal viscosity change at 1200°C, $\eta(T)$ = isobaric viscosity change at 3 GPa.

$\eta_0 = 6.5 \times 10^4$. Figure 7.1 illustrates the dominant control which temperature plays in determining mantle viscosity and indicates that the fluid dynamic behavior of the mantle wedge critically depends on the temperature distribution therein. The thermal structure of the mantle wedge has been calculated from surface heat flow data; however, temperature estimates vary widely at the surface of the subducting plate at 100-km depth, for example, from 400°C (Andrews and Sleep, 1974) to 1400°C (Griggs, 1972). Petrological data, especially the liquidus and solidus temperatures of basalts and peridotites, help out here for they provide key constraints upon the temperature existing at certain points in the mantle wedge.

7.1.2 Localized high temperatures (to ~1320°C) at ~30-km depth

It has been demonstrated, both from liquidus experiments and peridotite melting experiments, that primary low-K, volcanic front magmas in the NE Japan arc separate from the uppermost mantle wedge at 1300–1320°C and ~1.0 GPa (Chapter 5). These temperatures, at depths just beneath the crust/mantle boundary, are ~100°C and ~250°C higher than the solidus of anhydrous peridotite and basalt compositions, respectively. If temperatures >1300°C were present permanently near the Moho beneath the volcanic arc, large amounts of partial melting (possibly >20%; see Figure 3.17) might be expected to take place in the lowermost crust and uppermost mantle. This is unlikely, and is inconsistent with electrical conductivity and seismic data which show that partially molten regions are not pervasively distributed but spatially restricted to regions immediately beneath volcanoes (Chapter 2). High temperatures appear to be attained locally beneath each volcano as a result of ascent of mantle materials and/or partial melts from deeper and hotter parts of the mantle wedge.

7.1.3 Temperatures to ~1350°C in the mantle wedge

The temperature distribution and the maximum temperature attained in the mantle wedge are fundamental to understanding its rheology and dynamic behavior. Given adiabatic, isentropic ascent of partially melting mantle, the temperature at deeper levels of melting must be higher than at shallower levels, due to the energy (heat of fusion) consumed by the generation of melt (Figure 3.17). Adopting a heat of fusion of 200–400 J $kg^{-1} K^{-1}$ (McKenzie, 1984; Fukuyama, 1985), temperatures around 1350°C or more are necessary at deeper levels

(2–3 GPa) in the mantle wedge to produce the melt separation conditions of ~1320°C and 1.0 GPa (Tatsumi et al., 1983). These temperature estimates for the core of the mantle wedge are consistent with those expected for a mantle potential temperature of 1300 ± 20°C, though they are based strictly on anhydrous peridotite melting, since the heat of fusion for hydrous peridotite melting is not yet constrained. However, similar high-temperature estimates can be expected for hydrous mantle melting, because the conditions of melt separation for relatively hydrous backarc-side magmas in the NE Japan arc are 1320°C at 2.5 GPa.

The maximum temperatures estimated for the mantle wedge core are based on primary magma compositions estimated for the NE Japan arc whose MgO contents are 10–13 wt%. Equally, and in some cases even more, magnesian primary magmas, with up to 16 wt% MgO, have been constrained in other subduction zones (e.g., Aleutian Arc—Nye and Reid, 1986; Solomon Islands—Ramsay et al., 1984; Vanuatu—Barsdell, 1988; Barsdell and Berry, 1990; Eggins, 1993). These magnesian compositions require similar or perhaps even hotter mantle wedge core temperatures, given acceptable dissolved H_2O contents (i.e., <2–3 wt%).

7.1.4 Diapirs in the mantle wedge

Models of basalt magma genesis which involve diapiric ascent of partially molten mantle materials have been invoked by many petrologists (e.g., Green and Ringwood, 1967; Cawthorn, 1975; Marsh, 1979; Sakuyama, 1983b). Several lines of evidence are consistent with the occurrence of mantle diapirs in the mantle wedge: (1) spatially restricted occurrences of high-temperature (~1320°C) regions at ~30-km depth beneath the crust/mantle boundary, (2) the definition of low-V, probably high-T, columns beneath active volcanoes in the NE Japan arc (Hasemi et al., 1984; Zhao, 1991), (3) localized regions of shear-wave splitting and probable partial-melt presence beneath an active volcano in central Japan (Ando et al., 1983), and (4) distinct incompatible element characteristics of volcanoes in the NE Japan arc which are separated by distances as little as 10 km (Sakuyama, 1983b). The last of these observations suggests each volcano is sourced from a geochemically isolated region of the mantle wedge. This can be considered in terms of diapiric models for partial melting in the mantle wedge which may additionally account for deeper melt separation and smaller degrees of melting beneath the backarc-side volcanic chain (as discussed in Chapter 5) and the typical life span (10^4 to 10^5 years) of many individual composite volcanoes (Tomiya, 1991).

One counterargument against diapirism in the mantle wedge is the cooling which a diapir will undergo as it ascends into overlying cooler material. A mantle diapir develops and rises due to its buoyancy, with the velocity of ascent, V_s, being given by Stokes' law:

$$V_S = \frac{2}{3}\frac{\Delta\rho g R^2}{\eta_2}\left(\frac{\eta_2 + \eta_1}{2\eta_2 + 3\eta_1}\right) \approx \frac{\Delta\rho g R^2}{3\eta_2}$$

where η_1 and η_2 are the viscosities of the diapir and overlying mantle, $\Delta\rho$ is the density contrast, g is the acceleration due to gravity, and R is the diapir radius. The time for 40-km ascent is about 2 million years and V_s is calculated to be $\sim 2 \times 10^{-2}$ m yr^{-1}, for $\eta_2 = 10^{20}$ Pa·s, $\Delta\rho = 1 \times 10^2$ kg m^{-3}, and $R = 1.5 \times 10^4$ m. During the ascent period the diapir may cool by thermal diffusion. The governing equation for this process is

$$\frac{\partial T}{\partial t} = \kappa \frac{\partial^2 T}{\partial r^2} \qquad (0 \leqslant r < R)$$

where t is time, r is diapir radius, T is relative temperature ($T = 1$ at $0 \leqslant r < R$ and $T = 0$ at $r = R$ at $t = 0$), and κ is thermal diffusivity of mantle materials ($\sim 1 \times 10^{-6}$ m^{2}s^{-1}). Results calculated based on these equations are shown in Figure 7.2 for a diapir with a 15-km radius. They indicate the diapir will be cooled significantly, with its core temperature decreasing from 1400°C to ~1250°C, given an initial temperature for the surrounding mantle of 1200°C. This calculation does not take into account the reduced viscosity of the surrounding mantle due to heat loss from the diapir; however, Marsh (1978, 1979) and Marsh and Kantha (1978) have overcome this shortcoming by calculating the velocity of diapir ascent using the "hot Stokes" equation

$$V_h \approx \frac{\Delta\rho g R^2}{3\eta^*} Pe^{\frac{-3}{2}}$$

where η^* is the effective viscosity of the heated boundary layer surrounding the diapir and Pe is the Peclet number ($Pe = V_h R/\kappa$). Diapir velocity and the time to ascend 40 km are shown in Figure 7.3 as a function of effective viscosity. The boundary zone could have an effective viscosity $<10^{16}$ Pa·s if heating occurs to near solidus condi-

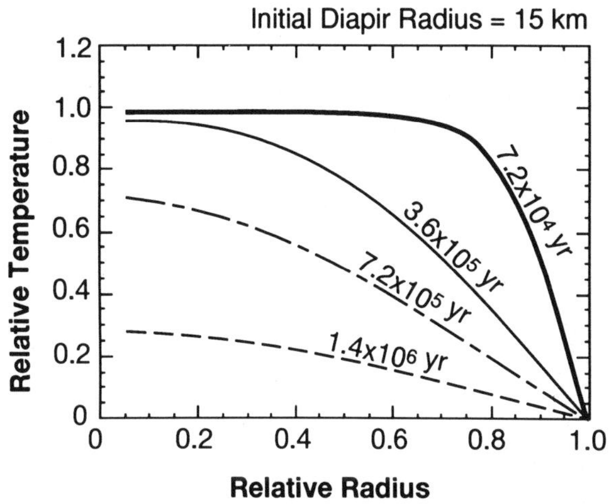

Figure 7.2 Radial cooling profiles for a mantle diapir with a 15-km radius as a function of time (and diapir ascent). Temperatures are given relative to the difference between the inital diapir temperature (1) and that of the surrounding mantle (0).

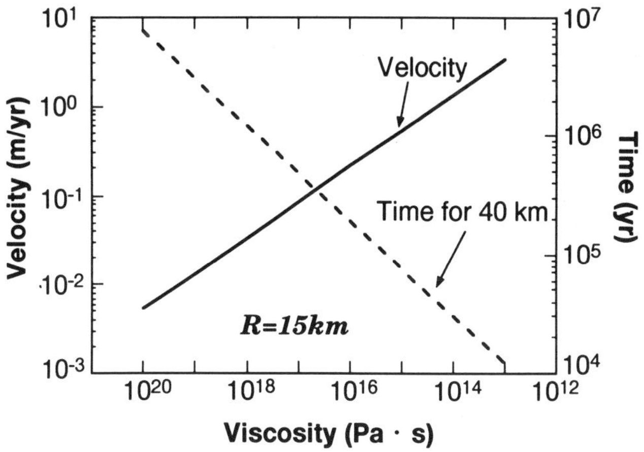

Figure 7.3 Calculated ascent velocity and time for 40-km rise of a mantle diapir with a 15-km radius as a function of effective viscosity of the surrounding mantle. Calculations made using the "hot stokes" equation.

tions, in which case cooling during diapiric ascent would be negligible. A diapir rising from the high-temperature (~1350°C) core of the mantle wedge at ~70 km may therefore reach a depth of ~30 km without appreciable cooling in a period of ~10^5 years (see also Figure 7.2).

Mantle diapirism and its role in basalt magma genesis have also been criticized from the point of view of melt segregation processes. Partially molten mantle will compact under gravity, expelling the less dense melt upward. It has been argued that this can take place at very small degrees of partial melting (<<2%; McKenzie, 1985), and hence it may be physically unrealistic for a mantle diapir to retain a large proportion of partial melt (i.e., up to ~20% for a low-K basalt magma). Consider the relative velocity (W_0) between melt and the solid matrix during buoyancy-driven porous flow, which is given by

$$W_0 = \frac{\kappa_\Phi (1-\Phi)(\rho_S - \rho_m) g}{\eta_m \Phi}$$

where κ_Φ is the permeability of the matrix, Φ is the porosity of melt fraction, ρ_s and ρ_m are the densities of the solid matrix and partial melt, g is the acceleration due to gravity, and η_m is the viscosity of the melt (McKenzie, 1985). The permeability may be expressed as

$$\kappa_\Phi = \alpha^2 \Phi^3 \times 10^{-3}$$

where a is the mean grain radius. Figure 7.4 shows variation in W_0 as a function of Φ for $\eta_m = 1$ Pa·s, $\Delta\rho = 5 \times 10^2\,\mathrm{kg\,m^{-3}}$, and $a = 10^{-3}$ m, which are appropriate for peridotite melting. If the degree of melting exceeds ~5%, the relative velocity of melt migration is much faster than the velocity of diapiric ascent, suggesting that porous flow rather than diapirism may dominate melt transport in the mantle wedge. However, we note that simple porous flow models may not apply to melt segregation within diapirs, as the stress environment and thermal gradient developed about the margins of a diapir are likely to retard melt separation so that relatively large melt porosities might develop.

The migration of partial melt within a melt-saturated, constant-porosity, porous, and deformable mantle wedge has been investigated by Spiegelman and McKenzie (1987). Given the assumptions of this model, melt flow is driven by pressure gradients formed in response to

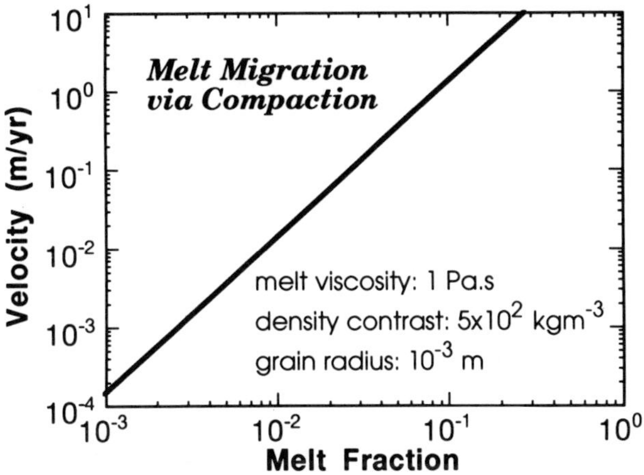

Figure 7.4 Example of calculated melt velocity relative to the solid matrix during upper mantle melting with melt segregation driven by compaction of the denser solid matrix.

viscous flow in the mantle wedge, which results in melt being drawn toward the mantle wedge corner at the junction between the volcanic arc lithosphere and the subducting plate. This model might apply if partial melt is homogeneously distributed throughout the mantle wedge, similar to that in the sub-mid-oceanic ridge mantle, and if the above solidus region is bounded by the surface of the subducting plate and volcanic arc lithosphere, but neither of these requirements seems likely. Another mode of melt migration based on buoyancy-driven porous flow, which has similarities to diapirs, has been proposed by Scott and Stevenson (1986). They show that for a system in which melt fraction changes little both temporally and spatially, melt flow concentrates into rising, high-porosity, spherical regions or "magmons." These are the three-dimensional equivalent of solitons or nonlinear solitary waves.

7.2 Flow in the Mantle Wedge

7.2.1 Subduction-induced convection in the mantle wedge

Subduction of the rigid oceanic plate into the viscous upper mantle will induce convection in the mantle wedge (Figure 7.5). Many models have been developed in order to investigate both the flow pattern and the influence of heat advection on the thermal structure in the mantle

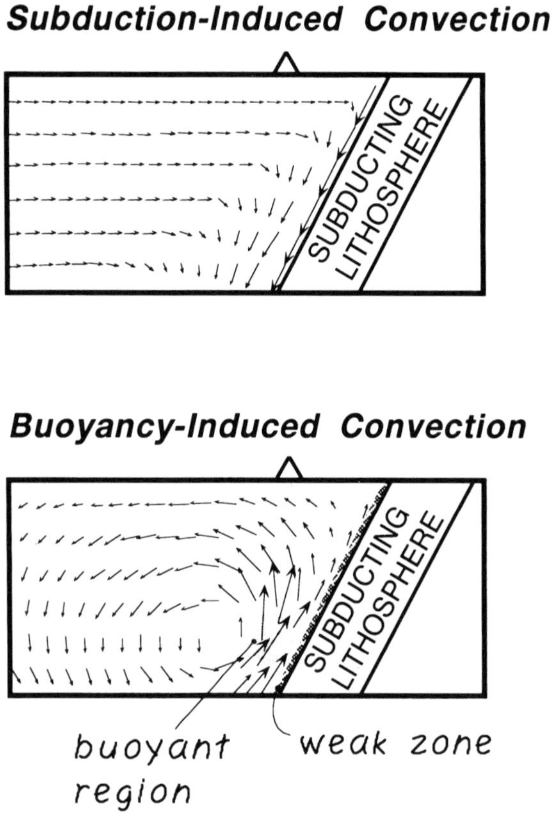

Figure 7.5 Two competing modes of convection in the mantle wedge, attributable to viscous coupling between the subducting slab and the mantle wedge and density difference between variably partially melted and unmolten mantle wedge materials.

wedge (McKenzie, 1969; Hasebe et al., 1970; Toksöz et al., 1971; Andrews and Sleep, 1974; Bodri and Bodri, 1978; Hsui et al., 1983; Honda, 1985; Peacock, 1990; Davies and Stevenson, 1992; Furukawa, 1993).

To simulate the subduction-induced convection requires solutions for the following: (1) the equation for conservation of mass or the equation of continuity, (2) the equation for conservation of momentum, and (3) the equation for conservation of energy. Although these have simple forms, the results of their application are highly variable due to several poorly constrained assumptions which need to be made.

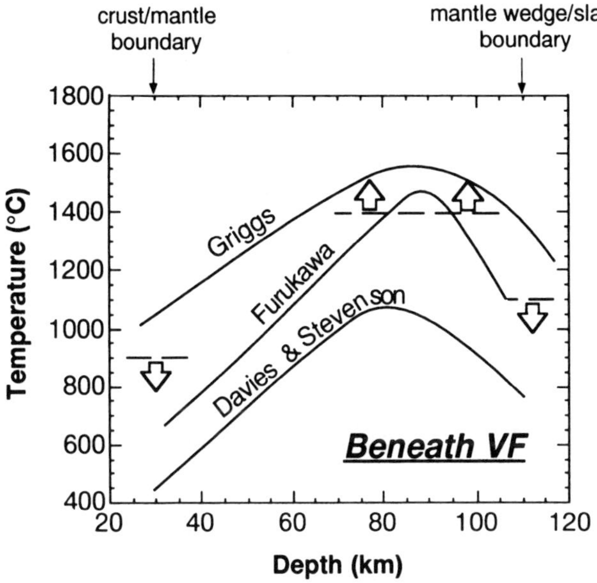

Figure 7.6 Examples of calculated temperature profile in the mantle wedge beneath the volcanic front based on models by Griggs (1972), Davies and Stevenson (1992), and Furukawa (1993). Petrological constraints applying to the thermal structure of the mantle are indicated by open arrows.

Several examples of published temperature profiles (geotherms) beneath the volcanic front, where the depth to the top of the subducted lithosphere is 110 km, are shown in Figure 7.6. These profiles vary greatly, and it is clear that tighter constraints are needed if more quantitative modeling is to be achieved. On the other hand, petrology is able to provide several external constraints which may be employed to refine these models, including (Figure 7.6) the following: (1) the temperature in the subducted oceanic crust must not exceed the anhydrous basalt solidus (~1100°C at 110 km) or else partial melting of the oceanic crust would occur; (2) the maximum temperature at the crust/mantle boundary must be <900°C at ~1.0 GPa or else the solidus of likely lower crustal materials (amphibolite or hydrous gabbro) would be exceeded; and (3) temperatures around 1350°C should be attained in the core of the mantle wedge in keeping with the melt generation conditions for subduction zone primary magmas. Two of the three model geotherms shown in Figure 7.6 do not meet or even approach these requirements. Models which attain very high temperatures have been based largely on the assumption of high temperatures in the original upper mantle. Most lower-temperature models have assumed constant viscosity in the mantle wedge, an assumption which requires the mantle wedge to continue to cool following the onset of subduction. Advocates of these models have invariably promoted the role of H_2O in the production of arc magmas, given the minimum temperature necessary to cause partial melting in the mantle wedge is ~1000°C (the solidus temperature of peridotite in the presence of excess H_2O).

A geothermal model that has incorporated the induced flow in the mantle wedge and has taken into account the petrological constraints outlined here was proposed for the NE Japan arc by Honda (1985). This model incorporated relatively high shear stress on the slab/wedge interface in order to satisfy surface heat flow data; however, Nagao and Uyeda (1989) and Furukawa and Uyeda (1989) have since demonstrated that there is substantial heat generation within the volcanic arc crust, and thus the need for lower shear stress on the thrust zone. Honda's model was criticized by Davies and Stevenson (1992) for placement of a high-temperature region in the mantle wedge beneath the backarc, but further developments by Furukawa (1993) have shown that the high-temperature mantle is preferentially transported toward the shallow mantle wedge corner by slab-induced convection, due to its low viscosity and ease of flow (Figure 7.7). This results in high temperatures in the mantle wedge near the wedge corner and beneath the volcanic front. Furthermore, if the temperature dependence of viscosity is taken into account, the effect of subduction velocity on the evolution of thermal structure may not be as significant as previously thought.

Numerical calculations of the subduction-induced convective flow have suggested the concentration of drag flow in the mantle wedge immediately above the surface of the subducting lithosphere (Figures 7.5 and 7.7). The hydrous peridotite layer developed in this region (Chapter 4) is likely to have a much lower viscosity than the overlying mantle wedge, due to the "lubricating" effect of the hydrous phases present such as serpentine and chlorite and due to the development of a shear zone near the surface of the subducting plate which may contribute additional substantial reduction in viscosity (Yuen, et al., 1978; Yuen and Schubert, 1979). It is probable that viscous coupling will be not be transmitted effectively through this hydrous peridotite layer into the overlying mantle wedge, and it may be reasonable to

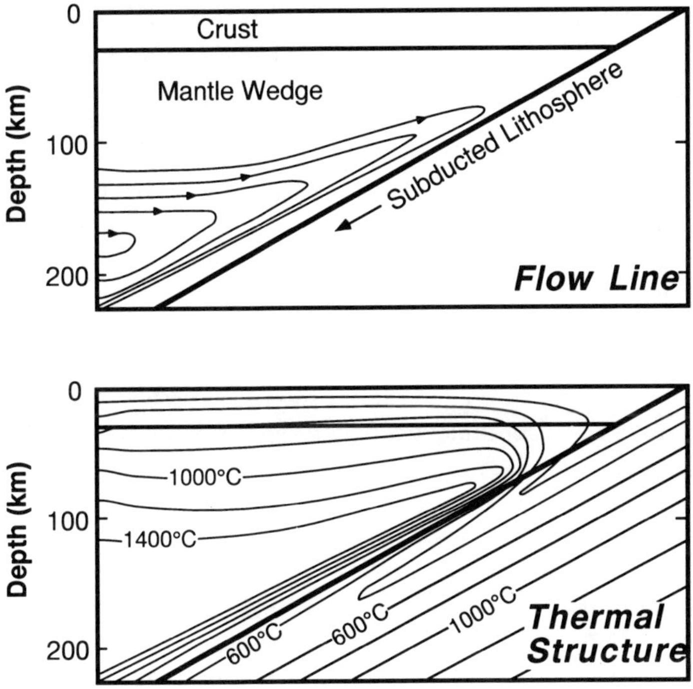

Figure 7.7 Convective flow lines for the mantle wedge induced by viscous coupling between the mantle wedge and the subducting slab, and a proposed thermal structure for the mantle wedge by Furukawa (1993).

assume that the width of this layer is ~10 km, a value similar to the width of the low-velocity layer documented by Matsuzawa et al. (1986) (Chapter 2). The approximate volume of material transported by the subduction-induced flow in the mantle wedge may be estimated accordingly to be 5×10^5 km^3/10^3 km Ma (10^3 km being a typical arc-scale length), assuming the velocity of the induced flow is ~5 cm/yr or half the typical rate of plate convergence in subduction zones.

7.2.2 Melt-buoyancy-driven convection and mantle wedge flow reversal

Partially molten materials will be less dense than the overlying/surrounding unmolten materials, and so trigger buoyant ascent of partially molten diapirs as discussed earlier. The direction of material transport associated with this buoyancy is essentially the reverse of that caused by subduction-induced convection (Figure 7.5). This flow reversal has the potential to play a significant role in material transport within the mantle wedge. The following assumptions, given that magmatism is facilitated by diaprism, provide a means for establishing an estimate of the flow reversal flux: (1) the volume of volcanic materials, excluding caldera-forming, possibly crust-derived felsic rocks, is ~$2 \times 10^3\,km^3/10^3\,km$ and ~$5 \times 10^2\,km^3/10^3\,km$ for the trench- and backarc-side volcanism in the NE Japan arc during the Quaternary (~2 Ma); (2) primary magmas undergo ~50% of fractional crystallization prior to eruption as typical andesite compositions; (3) trench- and backarc-side magmatism originate by 20% and 5% partial melting, respectively. The resultant volume estimate of $2 \times 10^4\,km^3/10^3\,km$ Ma for the material transportation by flow reversal is one order of magnitude smaller than the volume flux associated with subduction-induced convection. We conclude that flow reversal may readily take place in the core and upper parts of the mantle wedge, given that any opposing subduction-induced mantle wedge flow is likely to be concentrated in a thin layer along the base of the mantle wedge.

Flow reversal in the mantle wedge in the form of a continuous updraft of partially molten materials, rather than pulsed diapiric ascent, was proposed by Ida (1983), who suggested this form of flow reversal may predominate if the viscosity of the mantle wedge is less than 10^{20} Pa·s. However, Davies and Stevenson (1992) have since demonstrated that a viscosity smaller than 6×10^{18} Pa·s is required to produce this type of flow reversal, a value which is probably too small for the viscosity in the mantle wedge.

7.2.3 Mantle flow and magma production in the mantle wedge

We propose a schematic flow pattern in the mantle wedge based on competition between the primary subduction-induced convective flow and secondary convection associated with diapiric flow reversal (Figure 7.8). The proposed flow pattern is consistent with the transport of high-temperature mantle material into the shallower subvolcanic portion of the mantle wedge and with seismological and petrological observations which indicate isotherms in the mantle wedge are inclined upward from the backarc toward the volcanic arc (Chapters 2 and 5). The highest-temperature region is concentrated along with the incoming subduction-induced flow and is focused upward to a depth approaching 70 km beneath the volcanic front, aided by the influence of the reversed flow there. If the high-temperature region were to extend pervasively to shallower levels, then extensive partial melting

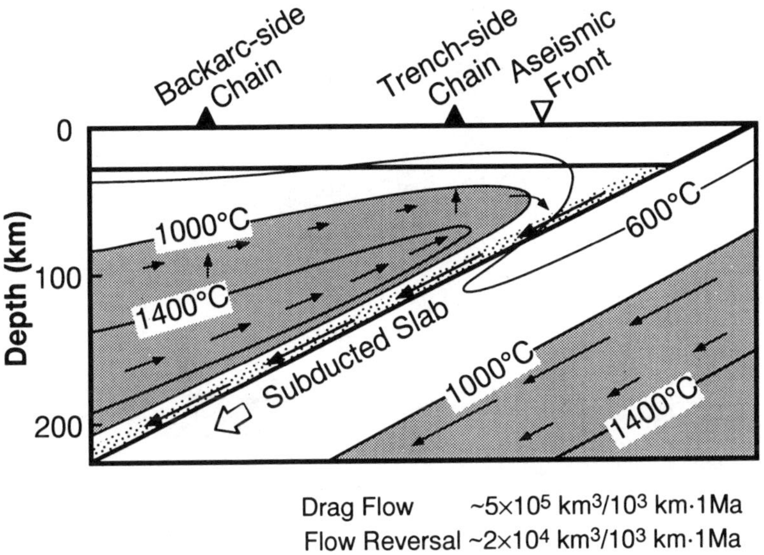

Figure 7.8 Proposed convective flow pattern (arrows) and thermal structure in the mantle wedge. Subduction-induced drag flow is limited to the base of the mantle wedge, and material transport by buoyancy-induced flow reversal is one order of magnitude smaller than that of the subduction-induced drag flow.

would take place, since 1350°C is the approximate solidus temperature of an anhydrous peridotite at 70-km depth. We also depict the concentration of the subduction-induced flow into a thin layer along the base of the mantle wedge.

Let us now trace the process of magma generation within the mantle wedge following formation of a hydrous peridotite layer which is dragged downward along the base of the mantle wedge (Figure 7.9). Aqueous fluid phases are liberated into the overlying mantle wedge from dehydration reactions involving amphibole/chlorite and phlogopite at ~3.5 GPa and ~6 GPa, respectively. These fluids migrate upward, halted by the crystallization of hydrous phases until successive addition of fluid results in H_2O-oversaturated conditions. Davies and Stevenson (1992) have suggested that the combination of vertical motion of the fluid phase and the transport of hydrous phases by the subduction-induced flow leads to a net horizontal transport of H_2O across the mantle wedge. However, if the downward subduction-induced flow in the mantle wedge is restricted to a thin layer at the

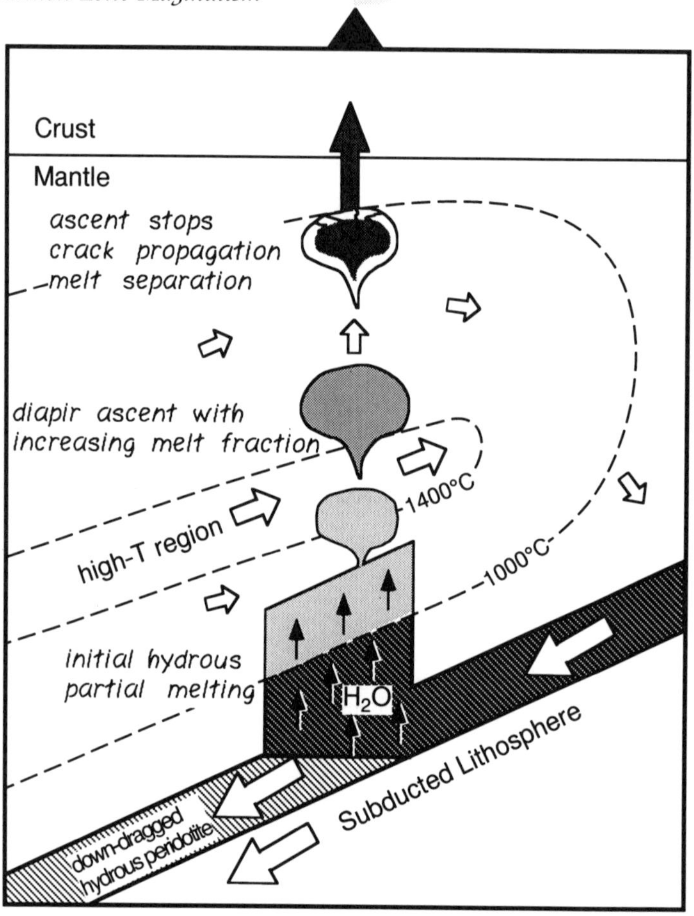

Figure 7.9 A schematic outline of the magma production process in subduction zones, in the context of solid (open arrows) and fluid (solid arrows) transport of material in the mantle wedge.

base of the mantle wedge, the horizontal component of H_2O transport is likely to be insignificant. Once the H_2O-fluid-oversaturated region, or hydrous curtain, extends into the mantle with a temperature exceeding ~1000°C, partial melting of mantle wedge peridotite occurs. Further supply of H_2O causes upward propagation of the partially molten region, with melt fraction increasing as the temperature increases toward the hot core of the mantle wedge. The degree of partial melting is initially low, and little relative melt migration through the solid matrix may take place. Buoyant partially molten diapirs, initiated from or above the region of incipient melting, ascend into the overlying core of the mantle wedge, heating as they rise and undergoing progressively greater extents of melting. Diapir ascent continues along with melting driven by decompression until halted near the base of the lithosphere, due to a decrease in temperature (to <~1000°C) and associated increase in viscosity of the surrounding wall rocks takes place (Figure 7.9). Melt separation accompanies cessation of diapir ascent and is facilitated by buoyant melt-filled crack propagation from the top of the diapir.

7.3 Deeper Processes in Subduction Zones

What is the fate of subducted oceanic lithosphere as it continues deeper into the mantle, beyond the relatively shallow region of magma generation restricted to the upper few hundred kilometers of the mantle? The fate of the subducting oceanic lithosphere is critical to understanding mantle convection and convective mixing processes, and to the development and location of chemical heterogeneity in the mantle, given that the subducted basaltic crust and sediments are so compositionally distinct from the mantle. It is an informative exercise to estimate roughly the amount of basaltic material which is subducted into the mantle. This turns out to be ~10^{10} km^3 over 10^9 years, based on the following uniformitarian assumptions: the mean rate of subduction is ~5 cm/yr over a total subduction zone length of ~4×10^4 km, and the thickness of the basalt layer is 5 km. This volume of basaltic material represents ~1% of the whole mantle, which if accumulated at the base of the mantle would form a chemically distinct layer ~100 km thick. Over the entire history of the Earth, present-day subduction rates would increase this estimate to ~5% of the mantle, and the total quantity of subducted material (including the lithospheric oceanic mantle) would account for ~50% of the mantle volume.

The greater overall density of the subducting oceanic lithosphere relative to that of the upper mantle is the driving force for subduction and a major driving or rather pulling component of the combined forces which give rise to surface plate tectonics (Forsyth and Uyeda, 1975). The density contrast between the oceanic lithosphere and surrounding upper mantle is a function of their differential temperature and their mineralogy. Understanding the phase changes and thus the mineralogy and density of basaltic and peridotitic compositions as a function of pressure and temperature is essential to our understanding of the fate of the subducted lithosphere. This will be investigated in some detail here, and some speculative extrapolations based on existing knowledge will be presented.

7.3.1 Phase relations of peridotite and basalt at ultrahigh pressures

It is now well accepted that the upper mantle is peridotitic. Pyrolite (*pyr*oxene-*ol*ivine-rock) is a model peridotite composition first proposed by Ringwood (1962a,b) and Green and Ringwood (1963), based on a mixture of basalt (20%) and refractory harzburgitic peridotite (80%). Pyrolite has a bulk composition close to that of the average of garnet peridotites from kimberlite pipes and consistent with that of fertile mantle peridotites which contain ~80% olivine and orthopyroxene (harzburgite), the remaining 20% comprising clinopyroxene and the aluminous phases plagioclase, spinel, and/or garnet (the basaltic component). Another proposed mantle rock type with an even greater basaltic component is piclogite (*pic*rite-ec*logite*), comprising less than 50% olivine and orthopyroxene and possessing a bulk composition between lherzolite and picrite (Anderson, 1979). Anderson and Bass (1986) have proposed that the mantle transition zone, between 400 and 650 km deep, is composed of piclogite, though this is not generally accepted. The composition of the lower mantle is the subject of even more debate, with proposals ranging from pyrolite/peridotite compositions (Ringwood, 1975; Jackson, 1983; Ito et al., 1984; Bukowinski and Wolf, 1990) to pure perovskite (the high-pressure polymorph of orthopyroxene). Some authors have suggested a change in chemical composition at the 650-km discontinuity to account for the change in geophysical properties at this the upper-lower mantle boundary, with a higher FeO and SiO_2 content in the lower mantle (Anderson, 1977, 1983; Anderson and Bass, 1986; Takahashi, 1986; Jeanloz and Knittle, 1989). However, consensus would currently have the lower mantle as being essentially peridotitic and close in composition to that of the upper mantle.

Since the early 1980s, uniaxial split-sphere multianvil apparatuses have been used to establish the phase relationships of both mantle and basaltic compositions to pressures of ~30 GPa, which correspond to the uppermost part of the lower mantle. Model mantle compositions, in both simple and chemically complex natural systems, have been examined mainly in Japan and Australia (Ito and Yamada, 1982; Ito et al., 1984; Irifune and Ringwood, 1987a,b; Ito and Takahashi, 1987; Ohtani and Sawamoto, 1987). The phase relationships of a

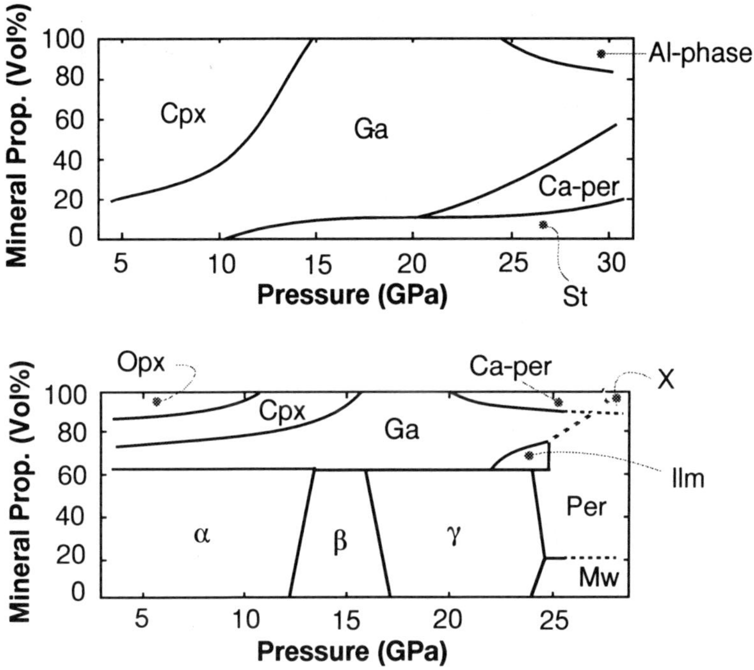

Figure 7.10 Phase relations of basalt (upper panel) and pyrolite (lower panel) compositions to ~30 GPa. Cpx = clinopyroxene, Ga = garnet, Per = perovskite, Opx = orthopyroxene, Ilm = ilmenite, Mw = Magnesiowüstite. *(Reproduced by permission from Irifune and Ringwood, 1987a, 1993.)*

pyrolite composition at 1200°C, determined by Irifune and Ringwood (1987a), are shown in modified form in Figure 7.10. Olivine (α) can be seen to change to spinel (β) at ~12 GPa, to spinel (γ) at ~16 GPa, and finally to perovskite and magnesiowüstite at ~25 GPa. Pyroxenes dissolve progressively into garnet with increasing pressure, until majorite garnet is the only phase existing other than olivine between 15 and 20 GPa. At pressures greater than 20 GPa, majorite garnet transforms to an assemblage including Ca-rich perovskite, ilmenite, and perovskite.

Seismic data have provided fundamental insights into and constraints upon the structure and physical properties of the Earth's interior. A seismological model of the mantle to 900 km deep is shown in Figure 7.11 based on the preliminary reference Earth model (PREM; Dziewonski and Anderson, 1981). The compressional- and shear-wave

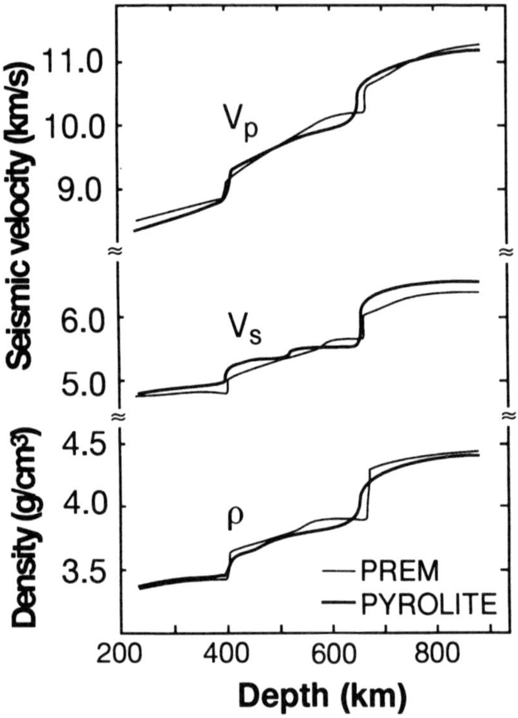

Figure 7.11 Calculated V_p, V_s, and density profiles for pyrolite after Irifune and Ringwood (1993) compared to the PREM model of Dziewonski and Anderson (1981). *(Reproduced by permission from Irifune and Ringwood, 1993.)*

velocities, V_p and V_s, are related to certain mineral physical properties as follows:

$$V_P = \left(\frac{K + \frac{4}{3}\mu}{\rho} \right)^{\frac{1}{2}}, \qquad V_S = \left(\frac{\mu}{\rho} \right)^{\frac{1}{2}}$$

where K, μ, and ρ are the bulk modulus, shear modulus, and density of the medium through which these seismic waves are transmitted. Given knowledge of these mineral physical properties, obtained from high-pressure experiments and by extrapolating results to conditions existing within the deep Earth interior (Lees et al., 1983; Weidner, 1985; Akaogi et al., 1987; Irifune, 1987), it is possible to assess plausible mantle compositions, mineralogy, and structure. Calculated V_p, V_s, and density as a function of depth are shown in Figure 7.11 (after Irifune and Ringwood, 1993), based on physical parameters given by Irifune (1987) and Irifune and Ringwood (1987a, 1993). The close comparison between the observed and calculated values for these parameters indicates that pyrolite is a viable mantle composition and that the 400- and 650-km discontinuities correspond to the sharp phase transitions from olivine to its modified spinel (β) polymorph and from spinel to perovskite and magnesiowüstite, respectively. It remains difficult to distinguish pyrolite and piclogite compositions in the depth range from 400 to 650 km; however, the bulk sound velocity variation across the 400-km discontinuity determined by Bina and Helffrich (1992) is best explained by an olivine content of ~70%, which is consistent with pyrolite/peridotite mantle at that depth. Whether the 650-km discontinuity is caused by an isochemical phase transformation of peridotite or by combined chemical and phase changes has also proved difficult to establish unequivocally.

Phase transformations in basalt compositions, particularly those of mid-oceanic ridge basalts (MORB) have also been studied extensively by Irifune et al. (1986) and Irifune and Ringwood (1987a, 1993). MORB compositions crystallize to an eclogite mineral assemblage (garnet and clinopyroxene) at pressures between ~1 and 10 GPa, with little change in the proportions of these phases occurring over this interval (Figure 7.10). At higher pressures, and by 15 GPa, clinopyroxene is dissolved into majorite garnet to form a garnetite assemblage (majorite garnet and relatively minor stishovite). Beyond 20 GPa, Ca-perovskite and an aluminous phase replace majorite garnet. The density variation occurring in the basaltic layer of the subducted lithosphere was calculated by Irifune and Ringwood (1993) based on these phase relationships and appropriate physical parameters and an appropriate geothermal model (e.g., Brown and Shankland, 1981) as shown in Figure 7.12. Ringwood (1990) noted that the oceanic basaltic crust may partially melt during subduction, though this would be unusual at least down to ~200-km depth beneath most volcanic arcs. If melting does occur, it will remove an SiO_2 component from the basaltic crust and result in a stishovite-free basaltic layer at higher pressures. In this case, the density of the subducted oceanic crust will be smaller than that shown in Figure 7.12 but will remain denser than mantle peridotite.

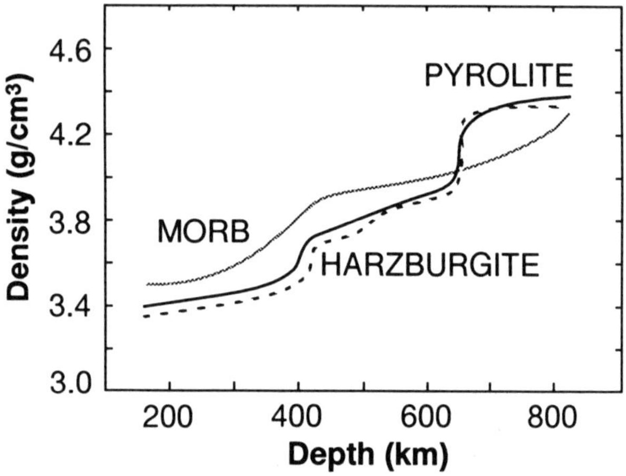

Figure 7.12 Comparison of the change in densities for components of the subducted oceanic lithosphere (MORB and harzburgite) and the mantle (pyrolite). *(Reproduced by permission from Irifune and Ringwood, 1993.)*

The calculated density change occurring with increasing depth within the uppermost harzburgitic layer of the oceanic lithosphere residual following MORB extraction beneath mid-ocean ridges is also shown in Figure 7.12.

7.3.2 Buoyancy relationships between subducting lithosphere and mantle

These differential density contrasts between the subducting oceanic lithosphere and the upper mantle through which it descends determine the magnitude of the forces driving subduction and influence the dynamic behavior of the subducting plate. Subducted oceanic crust of MORB composition is ~0.1–1.2 g/cm³ denser than pyrolitic peridotite throughout the entire upper mantle, and probably also at depths greater than 800 km within the lower mantle. This is due to the large proportion of garnet in the subducting crust, which is denser than olivine and its polymorphs at depths of <650 km. At depths between 650 and 800 km, basalt comprising the subducted oceanic crust is less dense than the surrounding lower mantle, which is due to a lower proportion of perovskite with a density greater than garnet. The harzburgitic part of the oceanic lithosphere is ~0.05–0.08 g/cm³ less dense than upper mantle pyrolite/peridotite to a depth of ~500 km, owing to its lower FeO content and smaller garnet component (lower Al_2O_3 content), and is 0.05 g/cm³ less dense than pyrolite/peridotite at depths below ~700 km, due to its lower FeO content and smaller perovskite component. This layer may be slightly denser than pyrolite/peridotite at depths between 650 and 700 km.

The principal driving force for subduction of oceanic lithosphere into the mantle is its overall greater density due to the relatively large contrast occurring between the basaltic layer and the upper mantle. However, the subducting oceanic lithosphere transfers to a region of positive buoyancy at the 650-km discontinuity (upper-lower mantle boundary), which extends to a depth of ~800 km. It follows that the subducted slab may be subjected to a resistive stress as it penetrates the uppermost part of the lower mantle.

Ringwood (1967) first suggested that the subducted basaltic oceanic crust might be trapped in a region of neutral buoyancy at the top of the 650-km discontinuity. He discussed the development of possible resistive stresses to subduction at the seismic discontinuities (Ringwood, 1975). Anderson (1979, 1980) speculated that the subducted basaltic crust is less dense than the surrounding mantle immediately below the 650-km discontinuity and that it will separate and accumulate there as a perched eclogite layer upon the upper/lower mantle boundary.

Possible stratification of the mantle has been investigated in considerable detail by Ringwood and co-workers (Ringwood, 1982, 1986, 1990, 1991; Ringwood and Irifune, 1988), who have proposed the formation of a melange or "megalith" of basaltic oceanic crust and depleted harzburgite which might attain a diameter of several hundred kilometers within the uppermost lower mantle (Figure 7.13). It was further suggested that the presence of the cold megalith may cool the underlying lower mantle and generate downwelling convective currents within the lower mantle, which may serve to erode and mix the megalith into the lower mantle. It has also been speculated that the basaltic portion of large megaliths may separate from the megalith and sink into the deep lower mantle in the form of a cold descending plume. On the other hand, if the subducted oceanic crust stagnates and thermally equilibrates with the mantle near the 650-km discontinuity, it may achieve neutral buoyancy and accumulate immediately above the discontinuity, without consequence for the lower mantle.

A number of numerical simulations of mantle convection have been made which take into account the phase transitions in the mantle and the subducted oceanic lithosphere (Machetel and Weber, 1991;

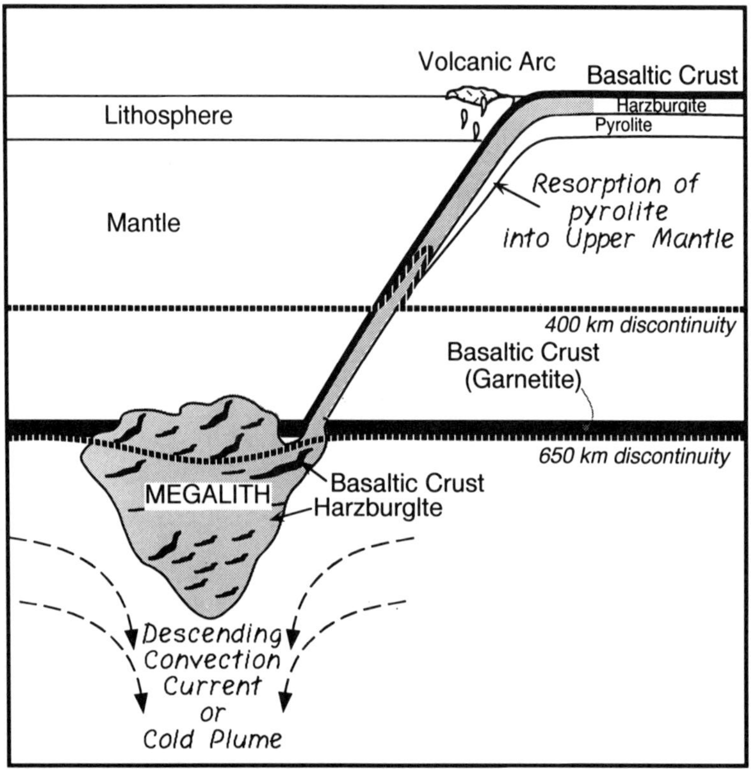

Figure 7.13 Formation of a megalith comprising the basaltic oceanic crust and the harzburgitic ultramafic layer of subducted lithosphere at the lower/upper mantle boundary, as proposed by Ringwood (1991).

Zhao et al., 1992; Honda et al., 1993). These illustrate some interesting effects which are broadly consistent with the concept of megalith formation. Cold sheets of both mantle and subducted oceanic crust materials may sink into the upper mantle but are halted temporarily at the 650-km discontinuity. The accumulated cold material ultimately gives rise to a gravitational instability that generates into a cold plume that descends rapidly into the lower mantle. These cold plumes may reach the core mantle boundary, where they are deflected laterally. This in turn displaces hot material from the layer immediately above the core-mantle boundary and may instigate the development of hot ascending plumes which may then continue to rise to the Earth's surface. It has been speculated on this basis that superplume events, manifested in the form of multiple large igneous provinces (oceanic plateaus and continental flood basalts), as have been observed between 120 and 75 Ma (Larson and Olson, 1991), might be attributed to hot ascending plume events triggered by a forerunner cold plume event.

7.4 Recycling of Subducted Oceanic Lithosphere Components

7.4.1 Ocean island basalts and hot mantle plumes

Hot-spot or ocean island basalt (OIB) volcanism is now widely believed to be a manifestation of hot plumes rising from the deep mantle, probably the thermal boundary layer surrounding the core (e.g., Campbell and Griffiths, 1990; Davies, 1990; Olson et al., 1990; Duncan and Richards, 1991). The "enriched" incompatible trace element and isotopic characteristics of OIB compared to MORB have been attributed to the accumulation of subducted MORB crust at a thermal boundary layer in the deep mantle (Hofmann and White, 1980, 1982; Davies, 1990). Moreover, the isotopic diversity observed in OIB, in particular the geochemically distinct end members that have become known as EM1, EM2, and HIMU (e.g., Hart, 1988), can be accounted for by the incorporation of different compositional components of subducted oceanic crust into their mantle sources. The EM1 and EM2 OIB end members have been argued to reflect the incorporation of distinct sediment compositions into the deep mantle via subduction, particularly EM2, which has very close similarities to crustally derived sediments (e.g., Weaver, 1991). The HIMU end member, on the other hand, possesses unusually radiogenic $^{206}Pb/^{204}Pb$ values ($^{206}Pb/^{204}Pb >$ 20.5; Hart, 1988) due to long-lived high-μ ($\mu = {}^{238}U/^{204}Pb$) values in its source materials, but is characterized by MORB-like $^{87}Sr/^{86}Sr$ and $^{143}Nd/^{144}Nd$ compared with other OIB end members, and has recently been shown to have high $^{187}Os/^{186}Os$ ratios compared with other OIB and MORB. Another feature of HIMU OIB is their relative enrichment of HFS elements, particularly of Nb and Ta in comparison to LIL elements (Dupuy et al., 1987; Hart, 1988; Weaver 1991; Chauvel et al., 1992). This is a reversal of the chemical fractionations observed in subduction zone lavas, which are attributed to the selective extraction of LIL elements from the subducted oceanic crust via subduction-derived fluids. This complimentarity in trace element geochemistry, combined with the Sr-, Nd-, and Os-isotope characteristics of HIMU

OIB strongly supports suggestions that the HIMU source constitutes a significant component of subducted basaltic oceanic crust which is returned to the Earth's surface via hot mantle plumes. The radiogenic Pb characteristics of HIMU OIB, requiring long-lived high-μ values, have been explained via preferential loss of Pb relative to U during dehydration of subducting oceanic crust at shallow mantle levels followed by long-term storage of the Pb-depleted crust in the deep mantle (Chauvel, 1992; Weaver, 1991). The model assumes that Pb is more soluble and readily transported than U during dehydration of the subducting oceanic crust. This appears to be consistent with the Pb-rich and low U/Pb ratios of subduction zone magmas, but has not yet been not confirmed by serpentine dehydration experiments (see Figure 6.4), in which U has been found to be more mobile than Pb. However, recent experimental results (Kogiso et al., unpublished data) indicated a higher degree of Pb transportation than U during amphibolite/eculgite transition; probably due to decomposition of Pb-rich sulfide minerals. Another possible mechanism producing the HIMU geochemical reservoir is the selective extraction of Pb relative to U into the metal core from the silicate mantle.

7.4.2 Other intraplate magmatism with possible subduction associations

A speculative scenario, based on the buoyancy relationships between subducted oceanic lithosphere and the upper mantle, is the possible formation of harzburgitic plumes rising from the transition zone at the base of the upper mantle (as proposed by Ringwood, 1982). If the harzburgite layer of the subducted lithosphere thermally equilibrates with the surrounding mantle, it will become buoyant and may form a rising diapir or plume. For example, if a volume of former harzburgitic oceanic lithosphere has a radius of 100 km and a density contrast of $-0.01\ g/cm^3$, relative to the upper mantle, then an ascent velocity of $\sim 10^2 m/yr$ will be achieved, given an upper mantle viscosity of 10^{18} Pa·s, and it would take $\sim 7 \times 10^3 yr$ for a plume of this material to ascend to the surface. Decompression melting of harzburgitic plume material might take place, given ascent to very shallow levels, or else more fertile mantle peridotite with a lower solidus temperature may be entrained and otherwise facilitate partial melting at deeper levels within the uppermost mantle. Alternatively, lithospheric mantle possessing a low solidus temperature might melt in response to conductive heating upon impingement of the plume upon the base of the lithosphere. In either case, partial melts might vary between the geochemical and isotopic signatures of the more fertile source materials and the depleted MORB-sourcelike harzburgitic plume material.

A possible case where this scenario has developed is in NE China, where extensive basaltic volcanism spanning tholeiitic and alkaline compositions has taken place during the Neogene (Figure 7.14). This volcanism is scattered irregularly in both space and time and covers an area more than 2000 km across. It does not seem likely that this volcanism reflects multiple hot-spot or hot mantle plume tracks, or a

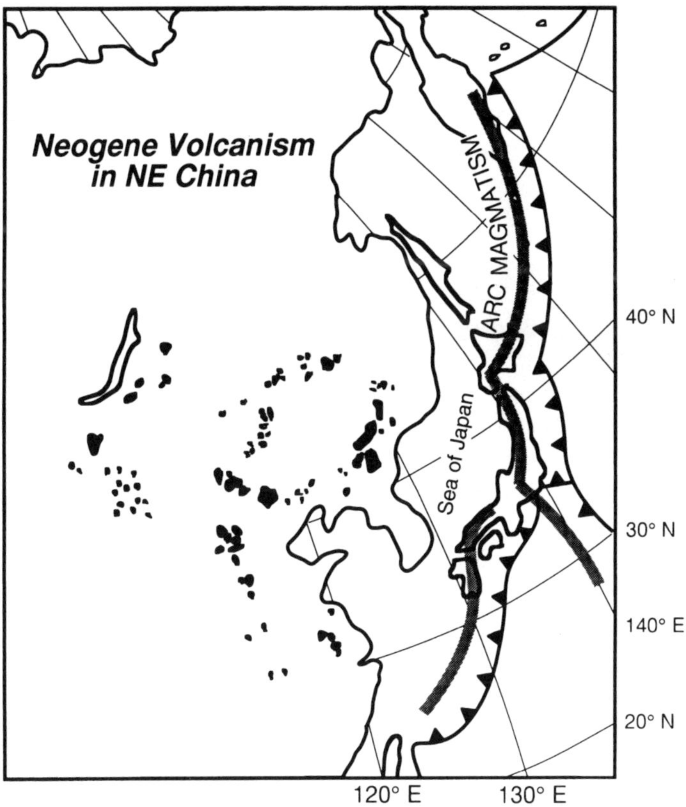

Figure 7.14 Cenozoic volcanic provinces in NE China, relative to subduction zones in the NW Pacific and position of the Japan Sea.

broad hot upwelling mantle region, as suggested by Miyashiro (1986), given the dimensions of typical hot spots and seismic tomography which does not indicate a thermal anomaly beneath the region (Fukao and Obayashi, 1992). Instead the tomographic images reveal a cold region formed at the base of the upper mantle and a restricted relatively high temperature region in the shallow upper mantle. These observations are consistent with the speculation that thermally neutral mantle material upwells beneath the region, driven by compositional density contrast alone, and that this material may derive from former harzburgitic layers of subducted lithosphere which have been stored near the lower/upper mantle boundary. Sr-Nd-isotopic data for NE China basalts (Li and Xue, 1983; Peng et al., 1986; Er and Zhao, 1987; Xie et al., 1988; Song and Frey, 1989) indicate that alkaline basalts have more depleted isotopic signatures, overlapping with those of MORBs, whereas tholeiitic basalts have more enriched and variable isotopic compositions. This relationship would require alkaline basalt magmas to be derived from the depleted harzburgitic source materials by small degrees of partial melting. This possibility requires further assessment.

7.4.3 Backarc rifting

Many volcanic arc-trench systems, especially those in the western Pacific, are associated with backarc basins (e.g., Taylor and Karner, 1983). The development of these basins is clearly associated with subduction processes, and various mechanisms have been proposed to explain their origin within this tectonic context (Elsasser, 1971; Toksöz and Bird, 1977; Chase, 1978; Molnar and Atwater, 1978; Uyeda and Kanamori, 1979; Dewey, 1980; Miyashiro, 1986). These mechanisms fall into two broad categories: "passive" processes, which emphasize the role of relative plate motions, and "active" processes, which invoke material flow within the mantle wedge to drive backarc extension.

One passive model, the "anchored slab" model, assumes that the locus of subduction is fixed and that the subducting lithosphere can migrate easily along its existing pathway but not move as a whole in a horizontal direction (Chase, 1978; Uyeda and Kanamori, 1979). Rifting of the overriding plate is constrained, if this were the case, to result from motion of the overriding lithosphere away from the trench position and the locus of subduction. Based on a fixed reference frame for plate motions, tied to hot-spot positions, Chase (1978) showed that the Philippine Sea plate behind the Mariana Trough has a component of motion away from the Izu-Mariana volcanic arc and trench axis of about 5 cm/yr. The model may also explain the absence of active backarc basins behind the Andes, Aleutian, and present-day NE Japan arc where the overriding plate motion is toward the trench axis. However, backarc rifting and basin development in the Scotia Sea and the Lau Basin are inconsistent with the operation of this mechanism, because the overriding plate has a trenchward component in both cases. The constraints placed on overriding plate motions are relaxed if the locus of subduction is able to migrate laterally. Molnar and Atwater (1978) proposed that sinking of the oceanic lithosphere under its own weight causes the trench to retreat oceanward, a process which has become known as *slab rollback*. Another possible cause of rollback is the oceanward buckling of the lithosphere that accommodates subduction of the spherical lithospheric plate (Yamaoka et al., 1986; Yamaoka and Fukao, 1987), similar to that undergone by the surface of a dented ping-pong ball. If the subducting plate is torn under such a situation, then the subducting plate may lose lateral support and undergo further "rollback" under its own weight. In support of this mechanism, Yamaoka et al., (1986) have demonstrated that most subduction zones with backarc basins are characterized by torn subducting plates. A similar mechanism, but causing the opposite case in which landward motion of the overriding lithosphere takes place, has been proposed by Kimura and Tamaki (1986), who speculated that backarc spreading in the Okhotsk and Japan seas was triggered by the landward migration of microplates as part of a large-scale deformation within the Eurasian plate resulting from the India-Eurasia collision (Tapponnier et al., 1982). The foregoing passive models for backarc rifting are confronted with some significant problems, including the transmission of forces between the subducting plate and the overriding lithosphere (e.g., Taylor and Karner, 1983) and, in particular, how to account for the observed change in stress orientation within the overriding lithosphere, with extension occurring in the backarc region and compression in the forearc region during backarc rifting and basin formation (Tatsumi and Tsunakawa, 1992).

An active backarc spreading model invoking high-temperature diapirs associated with subduction zone magmatism as the cause of extension has been proposed by Karig (1971, 1974). The difficulties with this model are (1) the limited occurrence of backarc rifting yet active magmatism and, by inference, mantle upwelling beneath all volcanic arcs, and (2) the location of some backarc spreading centers far from volcanic arcs (Taylor and Karner, 1983). Toksöz and Bird (1977) developed an earlier idea proposed by McKenzie (1969), emphasizing the role of subduction-induced convection in producing backarc rifting, whereby the lithosphere above the upward convective flow in the mantle wedge is stretched and rifted to form a backarc basin. If the coupling between the plates and convective flow is sufficient, then the resultant stress gradient may rotate from extension in the backarc region to compression nearer the trench (Tsunakawa, 1986). In this process, the dimensions of the convection cell within the mantle wedge are determined by the length of the subducting plate, with upwelling and resultant breakup of the lithosphere occurring above the leading tip of the downgoing plate. In most volcanic arc-trench systems associated with backarc basins, spreading centers are not located above the subducting plate edge but are nearer the trench axis, particularly during the initial stages of backarc rifting.

An active rifting model involving the forced injection of asthenospheric material into the mantle wedge was proposed by Miyashiro (1986) in the context of a model for upwelling "hot regions" behind western Pacific volcanic arcs with backarc basins. This model has been developed further by Tatsumi et al. (1989) and Tatsumi and Tsunakawa (1992) to explain the spatial and temporal variations in volcanism and orientation of the stress field in both the Miocene NE Japan arc and the present Taupo volcanic arc. Tectonic features common to these regions are (1) a trenchward shift of the volcanic front with time and a narrowing of the width of the volcanic arc immediately prior to the onset of backarc rifting, both of which suggest a steepening of the subducting plate, (2) variation in the stress gradient across these volcanic arcs, from extension in the backarc to compression in the forearc, and (3) the occurrence of "intraplate" volcanism in the far backarc region of both arcs: in NE China and in Northland, New Zealand. These characteristics, along with the occurrence of near-trench/forearc high-Mg andesite volcanism in the NE Japan arc during the Miocene, can be interpreted to result from the injection of high-temperature asthenospheric materials into the mantle wedge. It is proposed that the vertical mantle flow that causes intraplate volcanism in the far backarc region is deflected horizontally at the top of the asthenosphere, with a component toward the mantle wedge corner. This flow pushes both on the subducting plate and the rigid part of the mantle wedge corner beneath the forearc, and may also apply a significant drag force to the overriding plate. Tatsumi and Tsunakawa (1992) estimated the magnitude of these forces and concluded that the forces which act on the subducting plate and the rigid forearc mantle wedge predominate and are sufficient to cause backarc rifting and oceanward migration of the forearc and subducting plate. They are also consistent with the across-arc variation in stress gradient, from extension in the backarc to compression in the forearc.

Geochemical data for basalts from the backarc region of the NE Japan arc are consistent with the notion that the forced injection of mantle material with depleted geochemical characteristics into the mantle wedge may have induced backarc rifting. Secular variation in Sr-Nd-isotopic compositions of backarc-side magmas in the NE Japan arc (Nohda and Wasserburg, 1986; Nohda et al., 1988; Tatsumi et al., 1988) become more depleted with time, while trench-side magmas have retained rather constant Sr-Nd-isotopic compositions during the last 30 Ma (Figure 3.24). Basalt sills drilled at the Yamato Basin in the Japan Sea also show a similar secular variation to that of the backarc-side volcanic arc magmas (Nohda et al., 1992). These variations are reasonably explained by the replacement of preexisting isotopically enriched magma sources with depleted asthenospheric material. Yamashita and Tatsumi (1994) have estimated primary magma and their associated mantle source compositions for the Miocene NE Japan arc, and have argued that the magma source region beneath the backarc region of this arc comprises depleted harzburgitic mantle with a bulk composition similar to that remaining after extraction of N-MORB from a fertile upper mantle source. Based on this conclusion they speculated that the depleted magma source could be equated with the harzburgitic portion of the subducted lithosphere, which had been recycled into the mantle wedge by convective currents upwelling beneath NE China.

References

Acharya H. Volcanism and aseismic slip in subduction zones. J. Geophys. Res. 1981;86:335–344.

Akaogi M, Navrotsky A, Yagi T, Akimoto S. Pyroxene-garnet transformation: thermochemistry and elasticity of garnet solid solutions, and application to a pyrolite mantle. In: Manghnani M, Syono Y, eds. High-pressure research in mineral physics. Washington, DC: American Geophysical Union, 1987:251–260.

Aki K, Lee WHK. Determination of three-dimensional velocity anomalies under a seismic array using first P arrival times from local earthquakes. 1: a homogeneous initial model. J. Geophys. Res. 1976;81:4381–4399.

Allen JC, Boettcher AL. Amphiboles in andesite and basalt. II: Stability as a function of P-T-$f(H_2O)$-fO_2. Am. Mineral. 1978;60:1069–1087.

Anderson DL. Composition of the mantle and core. Ann. Rev. Earth Planet. Sci. 1977;5:179–202.

Anderson DL. The upper mantle: Eclogite? Geophys. Res. Lett. 1979;6:433–436.

Anderson DL. Chemical stratification of the mantle. J. Geophys. Res. 1980;84:6297–6298.

Anderson DL. Chemical composition of the mantle. J. Geophys. Res. (Suppl.) 1983;88:B41–B52.

Anderson DL, Bass JD. Transition region of the earth's upper mantle. Nature 1986;320:321–328.

Anderson OL, Grew PC. Stress corrosion theory of crack propagation with application to geophysics. Rev. Geophys. 1977;15:77–104.

Anderson RN, Uyeda S, Miyashiro A. Geophysical and geochemical constraints at converging plate boundaries. I: dehydration in the downgoing slab. Geophys. J. R. Astron. Soc. 1976;44:333–357.

Ando M, Ishikawa Y, Yamazaki F. Shear wave polarization anisotropy in the upper mantle beneath Honshu, Japan. J. Geophys. Res. 1983;88:5850–5864.

Andrews DJ, Sleep NH. Numerical modeling of tectonic flow behind island arcs. Geophys. J. R. Astron. Soc. 1974;38:237–251.

Ansell JH, Adams DA. Unfolding the Wadati-Benioff zone in the Kermadec–New Zealand Region. Phys. Earth Planet. Inter. 1986;44:274–280.

Aoki K, Ishikawa K, Kanisawa S. Fluorine geochemistry of basaltic rocks from continental and oceanic regions and petrogenetic application. Contrib. Mineral. Petrol. 1981;76:53–59.

Aramaki S, Ui T. Japan In: Thorpe RS, ed. Andesites. New York: Wiley, 1982:259–292.

Arculus RJ. Oxidation status of the mantle: past and present. Ann. Rev. Earth Planet. Sci. 1985;13:75–95.

Arculus RJ, Johnson RW. Criticism of generalized models for the magmatic evolution of arc-trench systems. Earth Planet. Sci. Lett. 1978;39:118–126.

Armstrong RL, Cooper JA. Lead isotopes in island arcs. Bull. Volcanol. 1971;35:27–63.

Ash RP, Carney JN, MacFarlane A. Geology of the Northern Bands Islands. Regional Rep. New Hebrides Govt. Geol. Survey 1980;52 pp.

Avdeiko GP, Volynets ON, Antonov AY, Tsvetkov AA. Kurile island-arc volcanism: structural and petrological aspects. Tectonophysics 1991;199:271–287.

Ayers JC, Watson EB. Rutile solubility and mobility in supercritical aqueous fluids. Contrib. Mineral. Petrol. 1993;114:321–330.

Bailey JC, Frolova TI, Burikova IA. Mineralogy, geochemistry and petrogenesis of Kurile island-arc basalts. Contrib. Mineral. Petrol. 1989;102:265–280.

Baker DR, Eggler DH. Fractionation paths of Atka (Aleutians) high-alumina basalts: constraints from phase relations. J. Volcanol. Geotherm. Res. 1983;18:387–404.

Baker MB, Newman S, Beckett JR, Stolper E. Separating liquid from crystals in high-pressure melting experiments using diamond aggregates. Geol. Soc. Am. Abstr. 1992.

Baker PE. The South Sandwich Islands. III: petrology of the volcanic rocks. Br. Antarctic Surv. Sci. Rep. 1978;93:1–34.

Ballhaus C. Redox states of lithospheric and asthenospheric upper mantle. Contrib. Mineral. Petrol. 1993;114:331–348.

Ballhaus C, Berry RF, Green DH. High pressure experimental calibration of the olivine-orthopyroxene-spinel oxygen barometer: implications for the oxidation state of the upper mantle. Contrib. Mineral. Petrol. 1991;107:27–40.

Bannister S. Microseismictiy and velocity structure in the Hawkes Bay region, New Zealand, fine structure of the subducting Pacific plate. Geophys. J. 1988;95:45–62.

Banno S, Matsui Y. On the formulation of partition coefficients for trace element distribution between minerals and magma. Chem. Geol. 1973;11:1–15.

Banno S, Sakai C, Higashino T. Pressure-temperature trajectory of the Sanbagawa metamorphism deduced from garnet zoning. Lithos. 1986;19:51–63.

Barazangi M, Isacks BL. Lateral variations of seismic-wave attenuation in the upper mantle above the inclined earthquake zone of the Tonga island arc: deep anomaly in the upper mantle. J. Geophys. Res. 1972;76:8493–8516.

Barazangi M, Isacks BL. Subduction of the Nazca plate beneath Peru: evidence from spatial distribution of earthquakes. Geophys. J. R. Astron. Soc. 1979;57:537–555.

Barsdell M. Petrology and petrogenesis of clinopyroxene-rich tholeiitic lavas, Merelava volcano, Vanuatu. J. Petrol. 1988;29:927–964.

Barsdell M, Berry R. The petrology and geochemistry of Western Epi. J. Petrol. 1990;31:747–777.

Barsdell M, Smith IEM, Sporli KB. The origin of reversed geochemical zoning in the northern New Hebrides volcanic arc. Contrib. Mineral. Petrol. 1982;81:148–155.

Basaltic Volcanism Study Project. Basaltic volcanism on the terrestrial planets. New York: Pergamon Press, 1981.

Basu AR, Tatsumoto M. Nd-isotopes in selected mantle-derived rocks and minerals and their implications for mantle evolution. Contrib. Mineral. Petrol. 1980;75:43–54.

Bebout GF, Ryan JG, Leeman WP. B-Be systematics in subduction-related metamorphic rocks: characterization of the subducted component. Geochem. Cosmochem. Acta 1993;57:2227–2237.

Benioff H. Orogenesis and deep crustal structure: additional evidence from seismology. Geol. Soc. Am. Bull. 1954;65:385–400.

Bina CR, Helffrich GR. Calculation of elastic properties from thermodynamic equation of state principles. Ann. Rev. Earth Planet. Sci. 1992;20:527–552.

Blackwell DD, Bowen RG, Hall DA, Riccio J, Steele JL. Heat flow, arc volcanism, and subduction in northern Oregon. J. Geophys. Res. 1982;87:8735–8754.

Bloomer SH. Goechemical characteristics of boninite- and tholeiite-series volcanic rocks from the Mariana forearc and the role of an incompatible element-enriched fluid in arc petrogenesis. Geol. Soc. Amer. Spec. Pap. 1987;215:151–164.

Bodri L, Bodri B. Numerical investigation of tectonic flow in island-arc areas. Tectonophysics 1978;50:163–175.

Bohlen SR, Boettcher AL, Wall VJ, Clemens JD. Stability of phlogopite-quartz and sanidine-quartz: a model for melting in the lower crust. Contrib. Mineral. Petrol. 1983;83:270–277.

Bowen NL. The evolution of the igneous rocks. Princeton: Princeton University Press, 1928.

Bowman JR, Ando M. Shear-wave splitting in the upper mantle wedge above the Tonga subduction zone. Geophys. J. R. Astron. Soc. 1987;88:25–41.

Brace WF, Byerlee JD. California earthquakes: why only shallow focus? Science 1970;168:1573–1575.

Brenan JM, Watson EB. Partitioning of trace elements between olivine and aqueous fluids at high P-T conditions: implications for the effect of fluid composition on trace-element transport. Earth Planet. Sci. Lett. 1991;107:672–688.

Brophy JG. The Cold Bay volcanic center, Aleutian volcanic arc. I: implications for the origin of hi-alumina arc basalt. Contrib. Mineral. Petrol. 1987;93:368–380.

Brophy JG, Marsh BD. On the origin of high-alumina arc basalt and the mechanics of melt extraction. J. Petrol. 1986;27:763–789.

Brown JM, Shankland TJ. Thermodynamic properties in the Earth as determined from seismic profiles. Geophys. J. R. Astron. Soc. 1981;66:579–596.

Brown L, Klein J, Middleton R, Sacks IS, Tera T. ^{10}Be in island-arc volcanoes and implications for subduction. Nature 1982;299:718–720.

Bukowinski M, Wolf G. Thermodynamically consistent decompression: implications for lower mantle composition. J. Geophys. Res. 1990;95:12583–12593.

Bulanova GP, Varshavskiy AV, Leskova NV, Nikishova LV. Central inclusions in natural diamonds. Dokl. Akad. Nauk SSSR 1979;244:704–706.

Burnham CW, Jahns RH. A method for determining the solubility of water in silicate melts. Am. J. Sci. 1962;260:721–745.

Cahill H, Isacks B. Seismicity and shape of the subducted Nazca plate. J. Geophys. Res. 1992;97:17503–17529.

Cameron WE. Contrasting boninite-tholeiite associations from New Caledonia. In: Crawford AJ, ed. Boninites. London: Unwin Hyman, 1989:314–338.

Campbell IH, Griffiths RW. Implications of mantle plume structure for the evolution of flood basalts. Earth Planet. Sci. Lett. 1990;99:79–93.

Cardwell RK, Isacks BL, Karig DE. The spatial distribution of earthquakes, focal mechanism solutions, and plate boundaries in the Philippine and northeastern Indonesian Islands. In: Hayes DE, ed. The tectonic and geologic evolution of Southeast Asian seas and islands. Geophys. Monogr., 23, Am. Geophys. Union, 1980:1–25.

Carr MJ. Symmetrical and segmented variation of physical and geochemical characteristics of the Central American volcanic front. J. Volcanol. Geotherm. Res. 1984;20:231–252.

Carr MJ, Rose WI, Mayfield DG. Potassium content of lavas and depth to the seismic zone in Central America. J. Volcanol. Geotherm. Res. 1979;5:387–401.

Carr MJ, Rose WI, Stoiber RE. Central America. In: Thorpe RS, ed. Andesites: orogenic andesites and related rocks. New York: Wiley, 1982:149–166.

Cathles LM. The viscosity of the Earth's mantle. Princeton: Princeton University Press, 1975.

Cawthorn RG. Degrees of melting in mantle diapirs and the origin of ultrabasic liquids. Earth Planet. Sci. Lett. 1975;27:113–120.

Chase CG. Extension behind island arcs and motions relative to hot spots. J. Geophys. Res. 1978;83:5385–5387.

Chauvel C, Hofmann AW, Vidal P. HIMU-EM: The French Polynesian connection. Earth Planet. Sci. Lett. 1992;110:99–119.

Chiu J, Isacks BL, Cardwell RK. Propagation of high-frequency seismic waves inside the subducted lithosphere from intermediate-depth earthquakes recorded in the Vanuatu arc. J. Geophys. Res. 1985;90:12741–12754.

Chopin C. Coesite and pure pyrope in high-grade blueschists of the Western Alps: a first record and some consequences. Contrib. Mineral. Petrol. 1984;86:107–118.

Chopin C, Henry C, Michard A. Geology and petrology of the coesite-bearing terrain, Dora Maira Massif, Western Alps. Eur. J. Mineral. 1991;3:263–291.

Christensen NI, Salisbury MH. Structure and constitution of the lower oceanic crust. Rev. Geophys. Space Phys. 1975;13:57–86.

Coats RR. Magma type and crustal structure in the Aleutian arc. In: Crust of the Pacific Basin. Geophys. Monogr. 6, Am. Geophys. Union, 1962:92–109.

Cohen RS, O'Nions RK. Identification of recycled continental material in the mantle from Sr, Nd and Pb isotope investigations. Earth Planet. Sci. Lett. 1982;61:73–84.

Cole JW. Tonga–Kermadec–New Zealand. In: Thorpe RS, ed. Andesites. New York: Wiley, 1982:245–258.

Cole JW. Taupo-Rotorua Depression, an ensialic marginal basin of North Island, New Zealand. In: Kokelaar BP, Howells MF, eds. Marginal basin geology. Oxford: Blackwell, 1984:109–120.

Condie KC, Potts MJ. Calcalkaline volcanism and the thickness of the early Precambrian crust in North America. Can. J. Earth Sci. 1969;6:1179–1184.

Conrad WK, Kay RW. Ultramafic and mafic inclusions from Adak Island: crystallization history, and implications for the nature of primary magmas and crustal evolution in the Aleutian Arc. J. Petrol. 1984;25:88–125.

Crampin S. Seismic wave propagation through a cracked solid: polarization as a possible dilatancy diagnostic. Geophys. J. R. Astron. Soc. 1978;53:467–496.

Crawford AJ, Cameron WE, Keays RR. The association boninite-low-Ti andesite-tholeiite in the Heathcote Greenstone Belt, Victoria: ensimatic setting for the early Lachlan Foldbelt. Austr. J. Earth Sci. 1984;31:197–208.

Crawford AJ, Falloon TJ, Eggins S. The origin of island arc high-alumina basalts. Contrib. Mineral. Petrol. 1987;97:417–430.

Crawford AJ, Falloon TJ, Green DH. Classification, petrogenesis and tectonic setting of boninites. In: Crawford AJ, ed. Boninites. London: Unwin Hyman, 1989:1–49.

Creager KC, Boyd TM. The geometry of Aleutian subduction: three-dimensional kinematic flow model. J. Geophys. Res. 1991;96:2293–2307.

Creager KC, Jordan TH. Slab penetration into the lower mantle beneath the Mariana and other island arcs of the northwest Pacific. J. Geophys. Res. 1986;91:3573–3589.

Cunningham P, Roecker SW, Hatzfeld D. Three-dimensional P and S wave velocity structures of southern Peru and their tectonic implications. J. Geophys. Res. 1986;91:9517–9532.

Dallwitz WB, Green DH, Thompson JE. Clinoenstatite in a volcanic rock from the Cape Vogel area, Papua. J. Petrol. 1966;7:375–403.

Danyushevsky LV, Falloon TJ, Sobolev AV, Crawford AJ, Carroll M, Price RC. The H_2O content of basalt glasses from Southwest Pacific back-arc basins. Earth Planet. Sci. Lett. 1993;117:347–362.

Davidson JP. Isotopic and trace element constraints on the petrogenesis of subduction-related lavas from Martinique, Lesser Antilles. J. Geophys. Res. 1986;91:5943–5962.

Davidson JP. Crustal contamination versus subduction zone enrichment: examples from the Lesser Antilles and implications for mantle source compositions of island arc volcanic rocks. Geochim. Cosmochim. Acta 1987;51:2185–2198.

Davies GF. Mantle plumes, mantle stirring and hotspot chemistry. Earth Planet. Sci. Lett. 1990;99:94–109.

Davies JH, Bickle MJ. A Physical model for the volume and composition of melt produced by hydrous fluxing above subduction zones. Phil. Trans. R. Soc. London 1991;335:355–364.

Davies JH, Stevenson DJ. Physical model of source region of subduction zone volcanics. J. Geophys. Res. 1992;97:2037–2070.

DeBari SM, Coleman RG. Examination of the deep levels of an island arc: evidence from the Tonsina ultramafic-mafic assemblage, Tonsina, Alaska. J. Geophys. Res. 1989;94:4373–4391.

DeBari SM, Kay SM, Kay RW. Ultramafic xenoliths from Adagdak Volcano, Adak, Aleutian Islands, Alaska: deformed igneous cumulates from the Moho of an island arc. J. Geol. 1987;95:329–341.

DeBoer J, Odom LA, Ragland PC, Snider PG, Telford NR. The Bataan orogene: Eastward subduction, tectonic rotations, and volcanism in the western Pacific (Philippines). Tectonophys. 1980;67:251–282.

Defant MJ, DeBoer JD, Oles D. The western central Luzon volcanic arc, the Philippines: two arcs divided by rifting? Tectonophysics 1988;145:305–317.

Defant MJ, Jacques D, Maury RC, DeBoer JD, Joron J-L. Geochemistry and tectonic setting of the Luzon arc, Philippines. Geol. Soc. Am. Bull. 1989;101:663–672.

Delany JM, Helgeson HC. Calculation of the thermodynamic consequences of dehydration in subducting oceanic crust to 100kb and >800°C. Am. J. Sci. 1978;278:638–686.

de Silva SL, Francis PW. Volcanoes of the central Andes. Germany: Springer-Verlag Berlin Heidelberg, 1991:216.

De Paolo DJ, Wasserburg GJ. Petrogenetic mixing models and Nd-Sr isotopic patterns. Geochim. Cosmochim. Acta 1979:43:615–627.

Dewey JF. Episodicity, sequence and style at convergent plate boundaries. In: Strangeway D, ed. The continental crust and its mineral resources, Spec.

Rep. 20. Toronto: Geological Association of Canada, Toronto, Ont., 1980:553–574.

Dickinson WR. Potassium-depth (K-h) relations in continental margin and intraoceanic magmatic arcs. Geology 1975;3:53–56.

Draper DS, Johnston AD. Anhydrous PT phase relations of an Aleutian high-MgO basalt: an investigation of the role of olivine-liquid reaction in the generation of arc high-alumina basalts. Contrib. Mineral. Petrol. 1992;112:501–519.

Drummond MS, Defant MJ. A model for trondhjemite-tonalite-dacite genesis and crustal growth via slab melting: archean to modern comparisons. J. Geophys. Res. 1990;95:21503–21521.

Duncan RA, Richards MA. Hotspots, mantle plumes, flood basalts, and true polar wander. Rev. Geophys. 1991;29:31–50.

Dupuy C, Vidal P, Barsczus HG, Chauvel C. Origin of basalts from the Marquesas Archipelago (south central Pacific Ocean): isotope and trace element constraints. Earth Planet. Sci. Lett. 1987;82:145–152.

Dziewonski AM, Anderson DL. Preliminary reference Earth model. Phys. Earth Planet. Inter. 1981;25:297–356.

Eaton JP, Lee WHK, Pakiser LC. Use of micro-earthquakes in the study of the mechanics of earthquake generation along the San Andreas Fault in central California. Tectonophysics 1970;9:252–282.

Edwards C, Menzies M, Thirlwall M. Evidence from Muriah, Indonesia, for the interplay of supra-subduction zone and intraplate processes in the genesis of potassic alkaline magmas. J. Petrol. 1991;32:555–592.

Eggins SM. Origin and differentiation of picritic arc magmas, Ambae (Aoba), Vanuatu. Contrib. Mineral. Petrol. 1993;1993:114.

Eggler DH. Role of CO_2 in melting processes in the mantle. Carnegie Inst. Wash. Yearb. 1973;73:457–467.

Eggler DH. Effect of CO_2 on the melting of peridotite. Carnegie Inst. Wash. Yearb. 1974;73:215–224.

Eggler DH. Solubility of major and trace elements in mantle metasomatic fluids: experimental constraints. In: Menzies MA, Hawkesworth CJ, eds. Mantle metasomatism. London: Academic Press, 1987:21–41.

Ellam RM, Hawkesworth CJ, Elemental and isotopic variations in subduction related basalts: evidence for a three component model. Contrib. Mineral. Petrol. 1988;98:72–80.

Ellam RM, Hawkesworth CJ, Menzies MA, Rogers NW. The volcanism of southern Italy: role of subduction and the relationship between potassic and sodic alkaline volcanism. J. Geophys. Res. 1989;94:4589–4601.

Elsasser WM. Seafloor spreading as thermal convection. J. Geophys. Res. 1971;76:1101–1112.

Engdahl ER, Gubbins D. Simultaneous travel-time inversion for earthquake location and subduction zone structure in the central Aleutian Islands. J. Geophys. Res. 1987;92:13855–13862.

Engdahl ER, Scholz CH. A double Benioff zone beneath the central Aleutians: an unbending of the lithosphere. Geophys. Res. Lett. 1977;4:473–476.

Er ML, Zhao DS. Cenozoic basalts and their mantle derived xenoliths in eastern China. Beijin (in Chinese): Science Press, 1987.

Ewart A. Mineralogy and chemistry of modern orogenic lavas—some statistics and implications. Earth Planet. Sci. Lett. 1976;31:417–432.

Ewart A, Hawkesworth CJ. The pleistocene-recent Tonga-Kermadec arc lavas: interpretation of new isotopic and rare earth data in terms of a depleted mantle source model. J. Petrol. 1987;28:495–530.

Falloon TJ, Green DH. Anhydrous partial melting of MORB pyrolite and other peridotite compositions at 10 kb: implications for the origin of primitive MORB glasses. Mineral. Petrol. 1987;37:181–219.

Falloon TJ, Green DH. Anhydrous partial melting of peridotite from 8 to 35 kb and the petrogenesis of MORB. J. Petrol. 1988;379–414.

Falloon TJ, Green DH, Hatton CJ, Harris KL. Anhydrous partial melting of a fertile and depleted peridotite from 2 to 30 kb and application to basalt petrogenesis. J. Petrol. 1988;29:1257–1282.

Falloon TJ, Green DH, McCulloch MT. Petrogenesis of high-Mg and associated lavas from the north Tonga Trench. In: Crawford AJ, ed. Boninites. London: Unwin Hyman, 1989:357–395.

Fedotov SA. Mechanism of magma ascent and deep feeding channels of island arc volcanoes. Bull. Volcanol. 1975;39:241–254.

Foley S. High-pressure stability of the fluor- and hydroxy-endmembers of pargasite and k-richterite. Geochim. Cosmochim. Acta 1991;55:2689–2694.

Ford CE, Russell DG, Craven JA, Fisk MR. Olivine-liquid equilibria: Temperature, pressure and composition dependence of the crystal/liquid cation partition coefficients for Mg, Fe^{2+}, Ca and Mn. J. Petrol. 1983;24:256–265.

Forsyth DW. Fault plane solutions and tectonics of the South Atlantic and Scotia Sea. J. Geophys. Res. 1975;80:1429–1443.

Forsyth DW, Uyeda S. On the relative importance of driving forces of plate motion. Geophys. J. R. Astron. Soc. 1975;43:163–200.

Fujii T, Scarfe CM. Composition of liquid coexising with spinel lherzolite at 10 kbar and the genesis of MORBs. Contrib. Mineral. Petrol. 1985;90:18–28.

Fujitani T, Masda A. Light REE inclination and distance from volcanic front; a case of volcanic rocks in Northeastern Japan. Geochem. J. 1981;15:269–281.

Fukao Y, Hori S, Ukawa M. A seismological constraint on the depth of basalt-eclogite transition in a subducting oceanic crust. Nature 1983;303:413–415.

Fukao Y, Obayashi M. Subducting slabs stagnant in the mantle transition zone. J. Geophys. Res. 1992;97:4809–4822.

Fukuyama H. Heat of fusion of basaltic magma. Earth Planet. Sci. Lett. 1985;73:407–414.

Furukawa Y. Magmatic processes under arcs and formation of the volcanic front. J. Geophys. Res. 1993;98:8309–8319.

Furukawa Y, Uyeda S. Thermal state under the Tohoku arc with consideration of crustal heat generation. Tectonophysics 1989;164:175–187.

Gaetani GA, Grove TL, Bryan WB. The influence of water on the petrogenesis of subduction-related igneous rocks. Nature 1993;365:332–334.

Gamble JA, Smith IEM, McCulloch MT, Graham IJ, Kokelaar BP. The geochemistry and petrogenesis of basalts from the Taupo volcanic zone and Kermadec island arc, S. W. Pacific. J. Volcanol. Geotherm. Res. 1993;54:265–290.

Gilbert MC, Helz RT, Popp RK, Spear FS. Experimental studies of amphibole stability. In: Veblen DR, Ribbe PH, eds. Reviews in mineralogy. 9B. Washington, DC: Mineralogical Society of America, 1982:229–353.

Gill JB. Orogenic andesite and plate tectonics. New York: Springer-Verlag, 1981.

Gill JB, Williams RW. Th isotopes and U-series studies of subduction related rocks. Geochim. Cosmochim. Acta 1990;54:1427–1442.

Goldschmidt VM. Geochemistry. Oxford: Oxford Univ. Press., 1954:730.

Gorton MP. The geochemistry and origin of Quaternary volcanism in the New Hebrides. Geochim. Cosmochim. Acta 1977;41:1257–1270.

Goto A, Tatsumi Y. Stability of chlorite in the upper mantle. Am. Mineral. 1990;75:105–108.

Goto K, Hamaguchi H, Suzuki Z. Earthquake generating stresses in a descending slab. Tectonophysics 1985;112:111–128.

Green DH. Experimental melting studies on a model upper mantle composition at high pressure under water-saturated and water-undersaturated conditions. Earth Planet. Sci. Lett. 1973;19:37–53.

Green TH, Green DH, Ringwood AE. The origin of high-alumina basalts and their relationships to quartz tholeiites and alkali basalts. Earth Planet. Sci. Lett. 1967;2:41–51.

Green TH, Pearson NJ. Ti-rich accessory phase saturation in hydrous mafic-felsic compositions at high P, T. Chem. Geol. 1986;54:185–201.

Green DH, Ringwood AE. Mineral assemblages in a model mantle composition. J. Geophys. Res. 1963;68:937–945.

Green DH, Ringwood AE. The genesis of basaltic magmas. Contrib. Mineral. Petrol. 1967;15:103–190.

Griggs DT. The sinking lithosphere and the focal mechanism of deep earthquakes. In: Robertson EC, ed. The nature of the solid Earth. New York: McGraw-Hill, 1972:361–384.

Guffanti M, Weaver CS. Distribution of late Cenozoic volcanic vents in the Cascade Range: volcanic arc segmentation and regional tectonic considerations. J. Geophys. Res. 1988;93:6513–6529.

Gust DA, Perfit MR. Phase relations of a high-Mg basalt from the Aleutian island arc: implications for primary island arc basalts and high-Al basalts. Contrib. Mineral. Petrol. 1987;97:7–18.

Haggerty SE. Redox state of the continental lithosphere. In: Menzies, MA, ed. Continental mantle. Oxford: Clarendon Press, 1990:87–109.

Haggerty SE, Tompkins LA. Redox state of Earth's upper mantle from kimberlitic ilmenites. Nature 1983;303:295–300.

Hamaguchi H, Goto K, Suzuki Z. Doubleplaned structure of intermediate-depth seismic zone and thermal stress in the descending plate. J. Phys. Earth. 1983;31:329–347.

Hamburger MW, Cardwell RK, Isacks BL. Seismotectonics of the northern Philippine island arc. In: Hayes D, ed. The tectonics and geological evolution of the Southeast Asian seas and islands, Part 2. Geophys. Monogr., 27, Am. Geophys. Union, 1982:1–22.

Hamilton DL, Burnham CW, Osborn EF. The solubility of water and effect of oxygen fugacity and water content on crystallization in mafic magmas. J. Petrol. 1964;5:21–39.

Hamilton W. Tectonics of the Indonesian region. US Geol. Surv. Prof. Pap. 1979;1078.

Hart SR. Heterogeneous mantle domains: signatures, genesis and mixing chronologies. Earth Planet. Sci. Lett. 1988;90:273–296.

Hart SR, Davis KE. Nickel partitioning between olivine and silicate melt. Earth. Planet. Sci. Lett. 1978;44:159–161.

Hasebe K, Fujii N, Uyeda S. Thermal processes under island arcs. Tectonophysics 1970;10:335–355.

Hasegawa A, Takagi A. Comparison of Wadati-Benioff zone geometry and distribution of earthquake generation stress beneath northeastern Japan and those beneath western South America. Sci. Rep. Tohoku Univ. Ser. 5 1987;31:1–18.

Hasegawa A, Umino N, Takagi A. Double-planed deep seismic zone and upper-mantle structure in the northeastern Japan Arc. Geophys. J. R. Astron. Soc. 1978b;54:281–296.

Hasegawa A, Umino N, Takagi A, Suzuki Z. Double-planed deep seismic zone and anomalous structure in the upper mantle beneath northeastern Honshu (Japan). Tectonophysics 1979;57:1–6.

Hasegawa A, Zhao D, Hori S, Yamamoto A, Horiuchi S. Deep structure of the northeastern Japan arc and its relationship to seismic and volcanic activity. Nature 1991;352:683–689.

Hasemi AH, Ishii H, Takagi A. Fine structure beneath the Tohoku district, northeastern Japan arc, as derived by an inversion of P-wave arrival times from local earthquakes. Tectonophysics 1984;101:245–265.

Hatherton T, Dickinson WR. Andesitic volcanism and seismicity in Indonesia, the Lesser Antiles, and other island arcs. J. Geophys. Res. 1969;74:5301–5310.

Hawkesworth CJ, Gallagher K, Hergt JM, McDermott F. Mantle and slab contributions in arc magmas. Ann. Rev. Earth Planet. Sci. 1993;21:175–204.

Hedge CE, Knight RJ. Lead and strontium isotopes in volcanic rocks from northern Honshu, Japan. Geochem. J. 1969;3:15–24.

Heming RF, Patterns of Quaternary basaltic volcanism in the northern North Island, New Zealand. N.Z. J. Geol. Geophys. 1980;23:335–344.

Henry SG, Pollack HN. Terrestrial heat flow above the Andean subduction zone in Bolivia and Peru. J. Geophys. Res. 1988;93:15153–15162.

Hickey RL, Frey FA, Gerlach DC. Multiple sources for basaltic rocks from the southern volcanic zone of the Andes (34°–41°S): trace element and isotopic evidence for contributions from subducted oceanic mantle and continental crust. J. Geophys. Res. 1986;91:5943–5962.

Hirahara K, Ikami A, Ishida M, Mikumo T. Three-dimensional P-wave velocity structure beneath central Japan: low-velocity bodies in the wedge portion of the upper mantle above high-velocity subducting plates. Tectonophysics 1989;163:63–73.

Hirose K, Kawamura K. A new experimental approach for incremental batch melting of peridotite at 15 kbar. Geophys. Res. Lett. 1994;21:2139–2142.

Hirose K, Kushiro I. Partial melting of dry peridotites at high pressures: determination of compositions of melts segregated from peridotite using aggregates of diamond. Earth Planet. Sci. Lett. 1993;114:477–489.

Hobbs BE, Ord A, Teyssier C. Earthquake in the ductile regime? Pure Appl. Geophys. 1986;124:309–336.

Hofmann AW, Feigenson MD. Case studies on the origin of basalt. I: theory and reassessment of Grenada basalts. Contrib. Mineral. Petrol. 1983;84:382–389.

Hofmann AW, White WM. The role of subducted oceanic crust in mantle evolution. Carnegie Inst. Wash. Year b. 1980;79:477–483.

Hofmann AW, White WM. Mantle plumes from ancient oceanic crust. Earth Planet. Sci. Lett. 1982;57:421–436.

Holloway JR. Volatile interactions in magmas. In: Newton RC, Navrotsky A, Wood BJ, eds. Thermodynamics of minerals and melts. 1. New York: Springer-Verlag, 1981:273–293.

Holloway JR, Lewis CF. CO_2 solubility in hydrous albite liquid at 5 kbar. Eos. Am. Geophys. Union Trans. 1974;55:483.

Holms A. Principles of physical geology. London and Edinburgh: Thomas Nelson and Sons, 1965:1022.

Honda S. Thermal structure beneath Tohoku, northeast Japan—a case study for understanding the detailed thermal structure of the subduction zone. Tectonophysics 1985;112:69–102.

Honda S, Yuen DA, Balachandar S, Reuteler D. Three-dimensional instabilities of mantle convection with multiple phase transitions. Science 1993;259:1308–1311.

Hori S. Seismic waves guided by untransformed oceanic crust subducting into the mantle: the case of the Kanto district, central Japan. Tectonophysics 1990;176:355–376.

Hori S, Inoue H, Fukao Y, Ukawa M. Seismic detection of the untransformed "basaltic" oceanic crust subducting into the mantle. Geophys. J. R. Astron. Soc. 1985;83:169–197.

Horiuchi S, Hasegawa A, Takagi A, Ito A, Suzuki M, Kameyama H. Mapping of a melting zone near Mt. Nikko-Shirane in northern Kanto, Japan, as inferred from S × P and S × S reflections. Sci. Rep. Tohoku Univ., Ser. 5 1988;31:43–55.

Horiuchi S, Ishii H, Takagi A. Two-dimensional depth structure of the crust beneath the Tohoku district, the northeastern Japan arc. I: method and Conrad discontinuity. J. Phys. Earth. 1982a;30:47–69.

Horiuchi S, Yamamoto A, Ueki A, Tachibana K, Kono T, Takagi A. Two-dimensional depth structure of the crust beneath the Tohoku district, the northeastern Japan arc. II: moho discontinuity and P-wave velocity. J. Phys. Earth 1982b;30:71–86.

Hsu LC. Selected phase relationships in the system Al-Mn-Fe-Si-O-H: a model for garnet equilibria. J. Petrol. 1968;9:40–83.

Hsui AT, Marsh BD, Toksöz MN. On the melting of the subducted oceanic crust: effects of subduction induced mantle flow. Tectonophysics 1983;99:207–220.

Hurukawa N, Imoto M. P and S velocities in the source region of subcrustal earthquakes in the Tokai district, central Japan. J. Phys. Earth 1987;35:1–17.

Hurukawa N, Imoto M. Subducting oceanic crusts of the Philippine Sea and Pacific plates. Geophys. J. Int. 1992;109:639–652.

Hutchison CS. Indonesia. In: Thorpe RS, ed. Andesites. New York: Wiley, 1982:207–224.

Ida Y. Convection in the mantle wedge above the slab and tectonic processes in subduction zones. J. Geophys. Res. 1983;88:7449–7456.

Inoue H, Fukao Y, Tanabe K, Ogata Y. Whole mantle P-wave travel time tomography. Phys. Earth Planet. Inter. 1990;59:294–328.

Irifune T. An experimental investigation of the pyroxene-garnet transformation in a pyrolite composition and its bearing on the constitution of the mantle. Phys. Earth Planet. Inter. 1987;45:324–336.

Irifune T, Ringwood AE. Phase transformations in primitive MORB and pyrolite compositions to 25 GPa and some geophysical implications. In: Manghnani M, Syono Y, eds. High pressure research in geophysics. Washington, DC: American Geophysical Union, 1987a:231–242.

Irifune T, Ringwood AE. Phase transformations in a harzburgite composition to 26 GPa: implications for dynamical behaviour of the subducting slab. Earth Planet. Sci. Lett. 1987b;86:365–376.

Irifune T, Ringwood AE. Phase transformations in subducted oceanic crust and buoyancy relationships at depths of 600–800 km in the mantle. Earth Planet. Sci. Lett. 1993;117:101–110.

Irifune T, Sekine T, Ringwood AE, Hibberson WO. The eclogite-garnetite transformations at high pressure and some geophysical implications. Earth Planet. Sci. Lett. 1986;77:245–256.

Irving AJ. A review of experimental studies of crystal/liquid trace element partitioning. Geochim. Cosmochim. Acta 1978;42:743–770.

Isacks B, Molnar P. Distribution of stresses in the descending lithosphere from a global survey of focal-mechanism solutions of mantle earthquakes. Rev. Geophys. Space Phys. 1971;9:103–174.

Isacks B, Oliver J, Sykes LR. Seismology and the new global tectonics. J. Geophys. Res. 1968;73:5855–5899.

Ishikawa K, Kanisawa S, Aoki K. Content and behavior of fluorine in Japanese Quaternary volcanoes of northeastern Japan. J. Volcanol. Geotherm. Res. 1980;8:161–175.

Ishikawa T, Nakamura E. Origin of the slab component in arc lavas from across-arc variation of B and Pb isotopes. Nature 1994;370:205–208.

Ito E, Takahashi E. Ultrahigh pressure phase transformations and the constitution of the deep mantle. In: Manghnani M, Syono Y, eds. High-pressure research in mineral physics. Washington, DC: American Geophysical Union, 1987:221–229.

Ito E, Takahashi E, Matsui Y. The mineralogy and chemistry of the lower mantle: an implication of the ultrahigh-pressure phase relations in the system MgO-FeO-SiO_2. Earth Planet. Sci. Lett. 1984;67:238–248.

Ito E, Yamada H. Stability relations of silicate spinels. ilmenites and perovskites. In: Akimoto S, Manghnani MH, eds. High-pressure research in geophysics. Tokyo: Terra Publ., 1982:405–419.

Iwamori H. Zonal structure of cenozoic basalts related to mantle upwelling in southwest Japan. J. Geophys. Res. 1991;96:6157–6170.

Iwamori H. Melt-solid flow with diffusion-controlled chemical reaction. Geophys. Res. Lett. 1992;19:309–312.

Jackson I. Some geophysical constraints on the chemical composition of the Earth's lower mantle. Earth Planet. Sci. Lett. 1983;62:91–103.

Jacob KH. Seismicity, tectonics, and geohazards of the Gulf of Alaska regions. In: Hood DW, Zimmerman ST, eds. The Gulf of Alaska, physical environment and biological resources. US Dept. of the Interior, Minerals Management Service, 1986:145–184.

Jakes P, Gill JB. Rare earth elements and the island arc tholeiitic series. Earth Planet. Sci. Lett. 1970;9:17–28.

Jakes P, White AJR. Structure of the Melanesian arcs and correlation with distribution of magma types. Tectonophysics 1969;8:223–236.

Jakes P, White AJR. K/Rb ratios of rocks from island arcs. Geochim. Cosmochim. Acta 1970;34:849–856.

Jakes P, White AJR. Major and trace element abundances in volcanic rocks of orogenic areas. Geol. Soc. Am. Bull. 1972;83:29–40.

Jaques AL, Green DH. Anhydrous melting of peridotite at 0–15 kbar pressure and the genesis of tholeiitic basalts. Contrib. Mineral. Petrol. 1980;73:287–310.

Jeanloz R, Knittle E. Density and composition of the lower mantle. Phil. Trans. R. Soc. London 1989; A328:377–389.

Jenner GA. Geochemistry of high-Mg andesites from Cape Vogel, Papua New Guinea. Chem. Geol. 1981;33:307–332.

Jezek P, Gill J, Whitford D, Duncan R. Reconnaissance geochemistry of the Sangihe-Sulawesi volcanic arc, Indonesia. Eos 1979;60:413.

Johnson KT, Kushiro I. Segregation of high pressure partial melts from peridotite using aggregates of diamond: a new experimental approach. Geophys. Res. Lett. 1992;19:1703–1706.

Johnson RW. Potassium variation across the New Britain volcanic arc. Earth Planet. Sci. Lett. 1976;31:184–191.

Johnson RW, Jaques AL, Hickey RL, McKee CO, Chappell BW. Manam Island, Papua New Guinea: petrology and geochemistry of a low-TiO_2 basaltic island-arc volcano. J. Petrol. 1985;26:283–323.

Johnston AD. Anhydrous P-T phase relations of near-primary high-alumina basalt form the South Sandwich Islands: implications for the origin of island arcs and tonalite-trondhjemite series rocks. Contrib. Mineral. Petrol. 1986;92:368–382.

Jordan TH. Lithospheric slab penetration into the lower mantle beneath the Sea of Okhotsk. J. Geophys. 1977;43:473–496.

Jordan TH, Lerner-Lam AL, Creager KC. Seismic imaging of boundary layers and deep mantle convection. In: Peltier WR, ed. Mantle convection, plate tectonics and global dynamics. New York: Gordon and Breach, 1989:97–201.

Kamiya S, Miyatake T, Hirahara K. How deep can we see the high velocity anomalies beneath the Japan Islands? Geophys. Res. Lett. 1988;15:828–831.

Kampfmann W, Berckhemer H. High temperature experiments on the elastic and anelastic behaviour of magmatic rocks. Phys. Earth Planet. Inter. 1985;40:223–247.

Karato S, Paterson MS, FitzGerald JD. Rheology of synthetic olivine aggregates: influence of grain size and water. J. Geophys. Res. 1986;91:8151–8176.

Karig DE. Origin and development of marginal basins in the western Pacific. J. Geophys. Res. 1971;76:2542–2561.

Karig DE. Evolution of arc systems in the western Pacific. Ann. Rev. Earth Planet. Sci. 1974;2:51–75.

Kawakatsu H. Double seismic zone in Tonga. Nature 1985;316:53–55.

Kawakatsu H. Double seismic zones: kinematics. J. Geophys. Res. 1986;91:4811–4825.

Kawano Y, Yagi K, Aoki K. Petrography and petrochemistry of the volcanic rocks of Quaternary volcanoes of northeastern Japan. Sci. Rep. Tohoku Univ. Ser. III 1961;7:1–46.

Kennedy WQ. Trends of differentiation in basaltic magmas. Am. J. Sci., Ser. 5 1933;25:239–256.

Kienle J, Swanson SE. Volcanism in the eastern Aleutian arc: late Quaternary and Holocene centers, tectonic setting and petrology. J. Volcanol. Geotherm. Res. 1983;17:393–432.

Kimura G, Tamaki K. Collision, rotation and back-arc spreading in the region of the Okhotsk and Japan seas. Tectonics 1986;5:389–401.

Kinzler RJ, Grove TL. Primary magmas of mid-ocean ridge basalts. 2: applications. J. Geophys. Res. 1992;97:6907–6926.

Kinzler BJ, Grove TL, Recca SI. An experimental study on the effect of temperature and melt composition on the partitioning of nickel between olivine and silicate melt. Geochim. Cosmochim. Acta 1990;54:1255–1265.

Kirby SH. Rheology of the lithosphere. Rev. Geophys. 1983;21:1458–1487.

Kitahara S, Takenouchi S, Kennedy GC. Phase relations in the system MgO-SiO_2-H_2O at high temperatures and pressures. Am. J. Sci. 1966;264:223–233.

Klein EM, Langmuir CH. Global correlations of ocean ridge basalt chemistry with axial depth and crustal thickness. J. Geophys. Res. 1987;92:8089–8115.

Kobayashi K, Nakada M. Magnetic anomalies and tectonic evolution of the Shikoku Inter-arc Basin. J. Phys. Earth 1978;26,Suppl.: S391–S402.

Koster van Groos AF, Wyllie PJ. Melting relationships in the system $NaAlSi_3O_8$-NaF-H_2O to 4kb pressure. J. Geol. 1968;76:50–70.

Kuge K, Satake K. Lateral segmentation within the subducting lithosphere: three-dimensional structure beneath the North Island, New Zealand. Tectonophysics 1987;139:223–237.

Kuno H. Origin of Cenozoic petrographic provinces of Japan and surrounding areas. Bull. Volcanol. 1959;20:37–76.

Kuno H. High-alumina basalt. J. Petrol. 1960;1:121–145.

Kuno H. Lateral variation of basalt magma type across continental margins and island arcs. Bull. Volcanol. 1966;29:195–222.

Kuroda N, Shiraki K, Urano H. Boninite as a possible calc-alkalic primary magma. Bull. Volcanol. 1978;41:563–575.

Kurtz RD, DeLaurier JM, Gupta JC. The electrical conductivity distribution beneath Vancouver Island: a region of active plate subduction. J. Geophys. Res. 1990;95:10929–10946.

Kushiro I. The system forsterite-diopside-silica with and without water at high pressures. Am. J. Sci. 1969;267-A:269–294.

Kushiro I. Effect of water on the composition of magmas formed at high pressures. J. Petrol. 1972;13:311–334.

Kushiro I. Melting of hydrous upper mantle and possible generation of andesite magma: an approach from synthetic systems. Earth Planet. Sci. Lett. 1974;22:294–299.

Kushiro I. On the lateral variations in chemical composition and volume of quaternary volcanic rocks across Japanese arcs. J. Volcanol. Geotherm. Res. 1983;18:435–447.

Kushiro I. A petrological model of the mantle wedge and lower crust in the Japanese island arcs. In: Mysen BO, ed. Geochem. Soc. Spec. Publ. 1987;1:165–181.

Kushiro I, Hirose K. Experimental determination of composition of melt formed by equilibrium partial melting of peridotite at high pressures using aggregates of diamond grains. Proc. Japan Acad. 1992;68:63–68.

Kushiro I, Sato H. Origin of some calc-alkalic andesites in Japanese islands. Bull. Volcanol. 1978;41:576–585.

Kushiro I, Schairer JF. New data on the system diopside-forsterite-silica. Carnegie Inst. Wash. Yearb. 1963;62:95–103.

Kushiro I, Shimizu N, Nakamura Y, Akimoto S. Compositions of coexisting liquid and solid phases formed upon melting of natural garnet and spinel lherzolites at high pressures: a preliminary report. Earth Planet. Sci. Lett. 1972;14:19–25.

Kushiro I, Syono Y, Akimoto S. Melting of a peridotite nodule at high pressures and high water pressures. J. Geophys. Res. 1968;73:6023–6029.

Lambert IB, Wyllie PJ. Melting of gabbro (quartz eclogite) with excess water to 35 kilobars, with geological applications. J. Geol. 1972;80:693–708.

Larson RL, Olson P. Mantle plumes control magnetic reversal frequency. Earth Planet. Sci. Lett. 1991;107:437–447.

LeBas MJ, LeMaitre RW, Streckeisen A, Zanettin B. A chemical classification of volcanic rocks based on the total alkali-silica diagram. J. Petrol. 1986;27:745–750.

Leeman WP. The influence of crustal structure on compositions of subduction-related magmas. J. Volcanol. Geotherm. Res. 1983;18:561–588.

Lees RC, Bukowinski MST, Jeanloz R. Reflection properties of phase transition and compositional change models of the 670 km discontinuity. J. Geophys. Res. 1983;88:8145–8159.

Lensen GJ, Otway PM. Earthshift and post-earthshift deformation associated with the May 1968 Inangahua earthquake, New Zealand. R. Soc. N. Z. Bull. 1971;9:107–116.

Li DS, Xue SH. The eastern China Mosozoic-Cenozoic basins and hydrocarbon occurrence. Acta Geol. Sinica 1983;3:224–233.

Lister JR, Kerr RC. Fluid-mechanical models of crack propagation and their application to magma transport in dykes. J. Geophys. Res. 1991;96:10049–10077.

Maaløe S, Wyllie PJ. Water content of a granite magma deduced from the sequence of crystallization determined experimentally with water-undersaturated conditions. Contrib. Mineral. Petrol. 1975;52:175–191.

Machetel P, Weber P. Intermittent layered convection in a normal mantle with an endothermic phase change at 670 km. Nature 1991;350:55–57.

Mallick DIJ. Some petrological and structural variations in the New Hebrides. In: Coleman PJ, ed. The western Pacific: island arcs, marginal seas, geochemistry. W. Austr. Univ. Press, 1973:193–212.

Marsh BD. Plume spacing and source. Nature 1975;256:240.

Marsh BD. On the cooling of ascending andesitic magma. Phil. Trans. R. Soc. London A 1978;288:611–625.

Marsh BD. Island arc development: some observations, experiments, and speculations. J. Geol. 1979;87:687–713.

Marsh BD. Magma chambers. Ann. Rev. Earth Planet. Sci. 1989;17:439–474.

Marsh BD, Carmichael ISE. Benioff zone magmatism. J. Geophys. Res. 1974;79:1196–1206.

Marsh BD, Kantha LH. On the heat and mass transfer from an ascending magma. Earth Planet. Sci. Lett. 1978;39:435–443.

Massonne H-J, Schreyer W. Stability field of the high-pressure assemblage talc + phengite and two new phengite barometers. Eur. J. Mineral. 1989;1:391–410.

Masuda Y, Nishimura S, Ikeda T, Katsui Y. Rare-earth and trace elements in the Quaternary volcanic rocks of Hokkaido, Japan. Chem. Geol. 1975;15:251–271.

Matsui Y, Onuma N, Nagasawa H, Higuchi H, Banno S. Crystal structure control in trace element partition between crystal and magma. Bull. Soc. Fr. Mineral. Cristallogr. 1977;100:315–324.

Matsuzawa T, Kono T, Hasegawa A, Takagi A. Subducting plate boundary beneath the northeastern Japan arc estimated from SP converted waves. Tectonophysics 1990;181:123–133.

Matsuzawa T, Umino N, Hasegawa A, Takagi A. Upper mantle velocity structure estimated from PS-converted wave beneath the north-eastern Japan arc. Geophys. J. R. Astron. Soc. 1986;86:767–787.

Matsuzawa T, Umino N, Hasegawa A, Takagi A. Estimation of thickness of a low velocity layer at the surface of the descending oceanic plate beneath the northeastern Japan arc by using synthesized PS-wave. Sci. Rep. Tohoku Univ., Ser. 5 1987;31:19–28.

McBirney AR. Andesitic and rhyolitic volcanism of orogenic belts. Am. Geophys. Union Monogr. 1969;13:501–506.

McBirney AR, Baker BH, Nilson RH. Liquid fractionation. I: basic principles and experimental simulations. J. Volcanol. Geotherm. Res. 1985;24:1–24.

McCaffrey R. Lithospheric deformation within the Molluca Sea arc-arc collision: evidence from shallow and intermediate earthquake activity. J. Geophys. Res. 1982;87:3663–3678.

McDermott F, Hawkesworth CJ. Th, Pb and Sr isotope variations in young arc volcanics and oceanic sediments. Earth Planet. Sci. Lett. 1991;104:1–15.

McKenzie DP. Speculations on the consequences and causes of plate motions. Geophys. J. R. Astron. Soc. 1969;18:1–32.

McKenzie D. The generation and compaction of partially molten rock. J. Petrol. 1984;25:713–765.

McKenzie D. The extraction of magma from the crust and mantle. Earth Planet. Sci. Lett. 1985;74:81–91.

McKenzie D, Bickle MJ. The volume and composition of melt generated by extension of lithosphere. J. Petrol. 1988;29:625–679.

McKenzie DP, Sclater JC. Heat flow in the eastern pacific and sea floor spreading. Bull. Volcanol. 1969;33:101–118.

Meen JK. Formation of shoshonites from calc-alkaline basalt magmas: geochemical and experimental constraints from the type locality. Contrib. Mineral. Petrol. 1987;97:333–351.

Meen JK. Elevation of potassium content of basaltic magma by fractional crystallization: the effect of pressure. Contrib. Mineral. Petrol. 1990;104:309–331.

Meijer A, Kwon TT, Tilton GR. U-Th-Pb partitioning behavior during partial melting in the upper mantle: implications for the origin of high Mucomponents and the "Pb Paradox." J. Geophys. Res. 1990;95:433–448.

Meissner R, Strehlau J. Limits of stresses in continental crusts and their relation to the depth-frequency distribution of shallow earthquakes. Tectonics 1982;1:73–89.

Menzies M, Murthy VR. Nd and Sr isotope geochemistry of hydrous mantle nodules and their host alkali basalts: implications for local heterogeneities in metasomatically veined mantle. Earth Planet. Sci. Lett. 1980;46:323–334.

Milhollen GL, Irving AJ, Wyllie PJ. Melting interval of peridotite with 5.7 percent water to 30 kilobars. J. Geol. 1974;82:575–587.

Minear JW, Toksöz MN. Thermal regime of a downgoing slab and new global tectonics. J. Geophys. Res. 1970;75:1397–1419.

Minster JF, Allegre CJ. Systematic use of trace elements in igneous processes. III: inverse problem of batch partial melting in volcanic suites. Contrib. Mineral. Petrol. 1978;68:37–52.

Mitronovas W, Isacks BL. Seismic velocity anomalies in the upper mantle beneath the Tonga-Kermadec island arc. J. Geophys. Res. 1971;76:7154–7180.

Mitropoulos P, Tarney J. Significance of mineral composition variations in the Aegean island arc. J. Volcan. Geotherm. Res. 1992;51:283–303.

Miyamachi M, Moriya T. Velocity structure beneath the Hidaka Mountains in Hokkaido, Japan. J. Phys. Earth 1984;32:13–42.

Miyashiro A. Metamorphism and metmorphic belts. London: Allen and Unwin, 1973.

Miyashiro A. Volcanic rock series in island arcs and active continental margins. Am. J. Sci. 1974;274:321–355.

Miyashiro A. Hot regions and the origin of marginal basins in the western pacific. Tectonophysics 1986;122:195–216.

Miyatake T, Hirahara K. Three-dimensional seismic velocity structure beneath the Japan Islands (3). Abstr. Fall Meeting of Seism. Soc. Japan, 1989:288.

Mizoue M, Nakamura I, Yokota T. Mapping of an unusual crustal discontinuity by microearthquake reflections in the earthquake swarm area near Asio, northwestern part of Tochigi Prefecture, central Japan. Bull. Earthq. Res. Inst. 1982;57:653–686.

Molnar P, Atwater T. Inter-arc spreading and cordilleran tectonics as alternatives related to the age of subducted oceanic lithosphere. Earth Planet. Sci. Lett. 1978;41:330–340.

Moore CB, Gibson EK, Larimer JW, Lewis CF, Nichiporuk W. Total carbon and nitrogen abundances in Apollo II lunar samples and selected chondrites and basalts. In: Proc. Apollo II Lunar Sci. Conf., 1, Suppl. 2. Geochim. Cosmochim. Acta, 1970:431–453.

Moore GF, Curray JR, Emmel FJ. Sedimentation in the Sunda trench and forearc region. Geol. Soc. London 1982; Spec. Paper 10:245–258.

Morris JD, Hart SR. Isotopic and incompatible element constraints on the genesis of island arc volcanics from Cold Bay and Amak Island, Aleutians and implications for mantle structure. Geochim. Cosmochim. Acta 1983;47:2015–2030.

Morris JD, Leeman WP, Tera F. The subducted component in island arc lavas: constraints from Be isotopes and B-Be systematics. Nature 1990;344;31–36.

Morris JD, Tera F. ^{10}Be and ^{9}Be in mineral separates and whole rocks from volcanic arcs: implication for sediment subduction. Geochim. Cosmochim. Acta 1989;53:3197–3206.

Mrozowski CL, Lewis SD, Hayes DE. Complexities in the tectonic evolution of the West Philippine Basin. Tectonophysics 1982;82:1–24.

Murase T, Fukuyama H. Shear wave velocity in partially molten peridotite at high pressures. Carnegie Inst. Wash. Yearb. 1980;79:307–310.

Murase T, Kushiro I. Compressional wave velocity in partially molten peridotite at high pressures. Carnegie Inst. Wash. Yearb. 1979;78:559–562.

Murton BJ, Peate DW, Arculus RJ, Pearce JA, van der Laan S. Trace-element geochemistry of volcanic rocks from site 786: the Izu-Bonin forearc In: Fryer P, Pearce JS, Stokking LR, eds. Proc. Ocean Drilling Program, Sci. Res. 125. TX: College Station, TX (Ocean Drilling Program), 1992:211–235.

Mysen BO, Boettcher AL. Melting of a hydrous mantle. I: phase relations of natural peridotite at high pressures and temperatures with controlled activities of water, carbon dioxide, and hydrogen. J. Petrol. 1975a;16:520–548.

Mysen BO, Boettcher AL. Melting of a hydrous mantle. II: geochemistry of crystals and liquids formed by anatexis of mantle peridotite at high pressures and high temperatures as a function of controlled activities of water, hydrogen, and carbon dioxide. J. Petrol. 1975b;16:549–590.

Mysen BO, Eggler DH, Seitz MG, Holloway JR. Carbon dioxide in silicate melts and crystals. I: solubility measurements. Am. J. Sci. 1976;276:455–479.

Mysen BO, Kushiro I. Compositional variations of coexisting phases with degree of melting of peridotite. Am. Mineral. 1977;62:843–865.

Nagao T, Uyeda S. Heat flow measurements in the northern part of Honshu, northeast Japan, using shallow holes. Tectonophysics 1989;164:301–314.

Nakagawa M. Geology and petrology of Moriyoshi volcano. J. Mineral. Petrol. Econ. Geol. 1983;78:197–210.

Nakamura K. Volcanoes as possible indicators of tectonic stress orientation—principle and proposal. J. Volcanol. Geotherm. Res. 1977;2:1–16.

Nakamura K, Uyeda S. Stress gradient in arc–back arc regions and plate subduction. J. Geophys. Res. 1980;85:6419–6428.

Nakamura Y, Kushiro I. Composition of the gas phase in Mg_2SiO_4-SiO_2-H_2O at 15 kbar. Carnegie Inst. Wash. Yearb. 1974;73:255–258.

Nakanishi I. Precursors to ScS phases and dipping interface in the upper mantle beneath southwestern Japan. Tectonophysics 1980;69:1–35.

Nakanishi I, Suyehiro K, Yokota T. Regional variations of amplitudes of ScSp phases observed in the Japanese Islands. Geophys. J. R. Astron. Soc. 1981;67:615–634.

Nicholls IA, Ringwood AE. Effect of water on olivine stability in tholeiites and production of silica-saturated magmas in the island arc environment. J. Geol. 1973;81:285–300.

Nicholls IA, Whitford DJ, Harris KL, Taylor SR. Variation in the geochemistry of mantle sources for tholeiitic and calc-alkaline mafic magmas. Western Sunda volcanic arc. Indonesia. Chem. Geol. 1980;30:177–199.

Nicolas A. A melt extraction model based on structural studies in mantle peridotites. J. Petrol. 1986;27:99–1022.

Nicolas A, Jackson M. High temperature dikes in peridotites: origin by hydraulic fracturing. J. Petrol. 1982;23:568–582.

Nockolds SR, Allen R. The geochemistry of some igneous rock series. Geochim. Cosmochim. Acta I 1953;4:105–142.

Nockolds SR, Allen R. The geochemistry of some igneous rock series. Geochim. Cosmochim. Acta II 1954;5:245–285.

Nockolds SR, Allen R. The geochemistry of some igneous rock series. Geochim. Cosmochim. Acta III 1956;9:34–77.

Nohda S, Tatsumi Y, Otofuji Y, Matsuda T, Ishizaka K. Asthenospheric injection and back-arc opening: isotopic evidence from NE Japan. Chem. Geol. 1988;68:317–327.

Nohda S, Tatsumi Y, Yamashita S, Fujii T. 57. Nd and Sr isotopic study of leg 127 basalts: implications for the evolution of the Japan sea backarc basin. Proc. Ocean Drilling Program, Sci. Res. 1992;127/128:899–904.

Nohda S, Wasserburg GJ. Nd and Sr isotopic study of volcanic rocks from Japan. Earth Planet. Sci. Lett. 1981;52:264–276.

Nohda S, Wasserburg GJ. Trends of Sr and Nd isotopes through time near the Japan Sea in northeastern Japan. Earth Planet. Sci. Lett. 1986;78:157–167.

Notsu K. Strontium isotope composition in volcanic rocks from the northeast Japan arc. J. Volcanol. Geotherm. Res. 1983;18:531–548.

Notsu K, Isshiki N, Hirano M. Comprehensive strontium isotope study of Quaternary volcanic rocks from the Izu-Ogasawara arc. Geochem. J. 1983;17:289–302.

Nye CJ, Reid MR. Geochemistry of primary and least fractionated lavas from Okmok volcano, central Aleutians: implications for arc magma genesis. J. Geophys. Res. 1986;91:10271–10287.

O'Hara MJ. Primary magmas and the origin of basalts. Scot. J. Geol. 1965;1:19–40.

Ohtani E, Sawamoto H. Melting experiment on a model chondritic mantle composition at 25 GPa. Geophys. Res. Lett. 1987;14:733–736.

Okada H. Forerunners of ScS waves from nearby deep earthquakes and upper mantle structure in Hokkaido. J. Seismol. Soc. Japan 1971;2–24:228–239.

Okada H. New evidences of the discontinuous structure of the descending lithosphere as revealed by ScSp phase. J. Phys. Earth 1979;27,Suppl.: S53–S63.

Okino K, Ando M, Kaneshima S, Hirahara K. The horizontally lying slab. Geophys. Res. Lett. 1989;16:1059–1062.

Oliver J, Isacks B. Deep earthquake zones, anomalous structures in the upper mantle, and the lithosphere. J. Geophys. Res. 1967;72:4259–4275.

Olson P, Silver PG, Carlson RW. The large-scale structure of convection in the Earth's mantle. Nature 1990;344:209–215.

Omura K, Kurita K, Kumazawa M. Experimental study of pressure dependence of electrical conductivity of olivine at high temperatures. Phys. Earth Planet. Interiors 1989;57:291–303.

Onuma N, Higuchi H, Wakita H, Nagasawa H. Trace element partition between two pyroxenes and the host lava. Earth Planet. Sci. Lett. 1968;5:47–51.

Oxtoby S, Hamilton DL. The discrete association of water with Na_2O and SiO_2 in NaAl silicate melts. Contrib. Mineral. Petrol. 1978;66:185–188.

Pacheco JF, Sykes LR, Scholz CH. Nature of seismic coupling along simple plate boudaries of the subduction type. J. Goephys. Res. 1993;98:14,133–14,159.

Page RW, Johnson RW. Strontium isotope ratios of Quaternary volcanic rocks from Papua New Guinea. Lithos 1974;7:91–100.

Pavoni N. Recent and late Cenozoic movements of the earth's crust. Roy. Soc. N. Z. Bull. 1971;9:7–17.

Peacock MA. Classification of igneous rock series. J. Geol. 1931;39:54–67.

Peacock SM. Numerical simulation of metamorphic pressure-temperature-time paths and fluid production in subducting slabs. Tectonics 1990;9:1197–1211.

Peng ZC, Zartman RE, Futa K, Chen DG. Pb, Sr and Nd isotopic systematics-chemical characteristics of Cenozoic basalts, eastern China. Chem. Geol. 1986;59:3–33.

Peppin WA, Exotic seismic phases recorded near Mammoth Lakes and their use in the delineation of shallow-crustal (Magma) anomalies. PAGEOPH 1987;125:1009–1023.

Peppin WA, Honjas W, Delaplain TW, Vetter UR. The case for a shallow-crustal anomalous zone (magma body?) near the south end of Hilton Creek Fault, California, including new evidence from an interpretation of pre-S arrivals. Bull. Seism. Soc. Am. 1989;79:805–812.

Perfit MR, Gust DA, Bence AE, Arculus RJ, Taylor SR. Chemical characteristics of island arc basalts: implications for mantle sources. Chem. Geol. 1980;30:227–256.

Plank T, Langmuir CH. An evaluation of the global variations in the major element chemistry of arc basalts. Earth Planet. Sci. Lett. 1993;90:349–370.

Powers HA. The lavas of the Modoc lava-bed Quadrangle, California. Am. Mineral. 1932;17:253–294.

Presnall DC, Dixon JR, O'Donnell TH, Dixon SA. Generation of mid-ocean ridge tholeiites. J. Petrol. 1979;20:3–36.

Ramsay WRH, Crawford AJ, Foden JD. Field setting, mineralogy, chemistry and genesis of arc picrites, New Georgia, Solomon Islands. Contrib. Mineral. Petrol. 1984;88:386–402.

Reyners M, Coles KS. Fine structure of the dipping seismic zone and subduction mechanics in the Shumagin Islands, Alaska. J. Geophys. Res. 1982;87:356–366.

Richter FM. Focal Mechanisms and seismic energy release of deep and intermediate earthquakes in the Tonga-Kermadec region and their bearing on the depth extent of mantle flow. J. Geophys. Res. 1979;84:6783–6795.

Riley JGN, Kohlstedt DL. Kinetics of melt migration in upper mantle-type rocks. Earth Planet. Sci. Lett. 1991;105:500–521.

Ringwood AE. A model for the upper mantle. J. Geophys. Res. 1962a;67:857–867.

Ringwood AE. A model for the upper mantle 2. J. Geophys. Res. 1962b;67:4473–4477.

Ringwood AE. The pyroxene-garnet transformation in the Earth's mantle. Earth Planet. Sci. Lett. 1967;2:255–263.

Ringwood AE. Composition and petrology of the Earth's mantle. New York: McGraw-Hill, 1975.

Ringwood AE. Petrogenesis in island arc systems. In: Talwani M, Pitman IWC, eds. Island arcs deep sea trenches and back-arc basins. Maurice Ewing Series 1. Washington, DC: Am. Geophys. Union, 1977:311–324.

Ringwood AE. Phase transformations and differentiation in subducted lithosphere: implications for mantle dynamics, basalt petrogenesis, and crustal evolution. J. Geol. 1982;90:611–643.

Ringwood AE. Dynamics of subducted lithosphere and implications for basalt petrogenesis. Terra Cognita 1986;6:67–77.

Ringwood AE. Slab-mantle interactions. 3: petrogenesis of intraplate magmas and structure of the upper mantle. Chem. Geol. 1990;82:187–207.

Ringwood AE. Phase transformations and their bearing on the constitution and dynamics of the mantle. Geochim. Cosmochim. Acta. 1991;55:2083–2110.

Ringwood AE and Green DH. An experimental investigation of the gabbro-eclogite transformation and some geophysical implications. Tectonophysics 1966;3:383–427.

Ringwood AE, Irifune T. Nature of the 650-km seismic discontinuity: implications for mantle dynamics and differentiation. Nature 1988;331:131–136.

Robertson JK, Wyllie PJ. Rock-water systems, with special reference to the water-deficient region. Am. J. Sci. 1971;271:252–277.

Roecker SW. Velocity structure of the Pamir-Hindu Kush region: possible evidence of subducted crust. J. Geophys. Res. 1982;87:945–959.

Roecker SW, Yeh YH, Tsai YB. Three-dimensional P and S wave velocity structures beneath Taiwan: deep structure beneath an arc-continent collision. J. Geophys. Res. 1987;92:10547–10570.

Roeder PL, Emslie RF. Olivine-liquid equilibrium. Contrib. Mineral. Petrol. 1970;29:275–289.

Ruff L, Kanamori K. Seismic coupling and uncoupling at subduction zones. Tectonophysics 1983;99:99–117.

Ryabchikov ID, Boettcher AL. Experimental evidence at high pressure for potassic metasomatism in the mantle of the Earth 1. Am. Mineral. 1980;65:915–919.

Ryan JG, Langmuir CH. The systematics of boron abundances in young volcanic rocks. Geochim. Cosmochim. Acta 1993;57:1489–1498.

Ryerson FJ, Watson EB. Rutile saturation in magmas: implications for Ti-Nb-Ta depletion in island arc basalts. Earth Planet. Sci. Lett. 1987;86:225–239.

Sacks IS. The subduction of young lithosphere. J. Geophys. Res. 1983;88:3355–3366.

Sacks IS, Okada H. A comparison of the anelasticity structure beneath western South America and Japan. Phys. Earth Planet. Inter. 1974;9:211–219.

Sakuyama M. Lateral variation of phenocryst assemblages in volcanic rocks of the Japanese Islands. Nature 1977;269:134.

Sakuyama M. Lateral variation of H_2O contents in Quaternary magmas of northeastern Japan. Earth Planet. Sci. Lett. 1979;43:103–111.

Sakuyama M. Petrological study of the Myoko and Kurohime volcanoes, Japan: crystallization sequence and evidence for magma mixing. J. Petrol. 1981;22:553–583.

Sakuyama M. Phenocryst assemblages and H_2O content in circum-Pacific arc magmas. In: Hilde TWC, Uyeda S, eds. Geodynamics of the western Pacific-Indonesian region. Geodynamic Series 11:1983a;143–157.

Sakuyama M. Petrology of arc volcanic rocks and their origin by mantle diapirs. J. Volcanol. Geotherm. Res. 1983b;18:297–320.

Sakuyama M. Magma mixing and magma plumbing systems in island arcs. Bull. Volcanol. 1984;47–4:685–703.

Sakuyama M, Kushiro I. Vesiculation of hydrous andesitic melt and transport of alkalis by separated vapor phase. Contrib. Mineral. Petrol. 1979;71:61–66.

Sakuyama M, Nesbitt RW. Geochemistry of the Quaternary volcanic rocks of the northeast Japan arc. J. Volcanol. Geotherm. Res. 1986;29:413–450.

Sanford AR, Alptekin O, Toppozada TR. Use of reflection phases on microearthquake seismograms to map an unusual discontinuity beneath the Rio Grande Rift. Bull. Seism. Soc. Am. 1973;63:2021–2034.

Sato H. Nickel content of basaltic magma: identification of primary magmas and measure of the degree of olivine fractionation. Lithos 1977;10:113–120.

Sato H. Thermal structure of the mantle wedge beneath northeastern Japan: magmatism in an islcal arc for the combined data of seismic anelasticity and velocity at heat-flow. J. Volcanol. Geotherm. Res. 1992;51:237–252.

Sato H, Sacks IS, Murase T, Muncill G, Fukuyama H. Qp-melting temperature relation in peridotite at high pressure and temperature: attenuation mechanism and implications for the mechanical properties of the upper mantle. J. Geophys. Res. 1989;94:10647–10661.

Sato H, Sacks IS, Takahashi E, Scarfe CM. Geotherms in the Pacific Ocean from laboratory and seismic attenuation studies. Nature 1988;336:154–156.

Saunders AD, Rogers G, Marriner GF, Terrell DJ, Verma SP. Geochemistry of Cenozoic volcanic rocks, Baja California, Mexico: implications for the petrogenesis of post-subduction magmas. J. Volcanol. Geotherm. Res. 1987;32:223–245.

Savage MK, Shih XP, Peyer RP, Aster RC. Shear-wave anisotropy of active tectonic regions via automated S-wave polarization analysis. Tectonophysics 1989;165:279–292.

Schmidt MW, Poli S. The stability of lawsonite and zoisite at high pressures: experiments in CASH to 92 kbar and implications for the presence of hydrous phases in subducted lithosphere. Earth Planet. Sci. Lett. 1994;124:105–118.

Schneider ME, Eggler DH. Fluids in equilibrium with peridotite minerals: implications for mantle metasomatism. Geochim. Cosmochim. Acta 1986;50:711–724.

Scott DR, Stevenson DJ. Magma ascent by porous flow. J. Geophys. Res. 1986;91:9238–9296.

Sekine T, Wyllie PJ. Phase relationships in the system $KAlSiO_4$-Mg_2SiO_4-SiO_2-H_2O as a model for hybridization between hydrous siliceous melts and peridotite. Contrib. Mineral. Petrol. 1982;79:368–374.

Selig F. A theoretical prediction of salt-dome patterns. Geophysics 1965;30:633–643.

Shankland TJ, O'Connell RJ, Waff H. Geophysical constraints on partial melt in the upper mantle. Rev. Geophys. Space. Phys. 1981;19:394–406.

Shaw DM. Trace element fractionation during anatexis. Geochim. Cosmochim. Acta 1970;34:237–243.

Shaw HR. The fracture mechanisms of magma transport from the mantle to the surface. In: Hargraves RB, ed. Physics of magmatic processes. Princeton: Princeton University Press, 1980:201–264.

Shih TC. Magnetic Lineations in the Shikoku Basin. In: Initial reports of Deep Sea Drilling Project. 58. Washington, DC: U.S. Government Printing Office, 1980:783–788.

Shimozuru D, Kubo N. Volcano spacing and subduction. In: Shimozuru D, Yokoyama I, eds. Arc volcanism: physics and tectonics. Tokyo: Terra Pub., 1983:141–151.

Shuto K. Miocene abyssal tholeiite-type basalts from Tobishima Island, eastern margin of the Japan Sea. J. Mineral. Petrol. Econ. Geol. 1988;83:257–272.

Sibson RH. Fault zone models, heat-flow, and the depth distribution of earthquakes in the continental crust of the United States. Bull. Seism. Soc. Am. 1982;72:151–163.

Sibson RH. Roughness at the base of the seismogenic zone: contributing factors. J. Geophys. Res. 1984;89:5791–5799.

Sigurdsson IA, Kamenetsky VS, Crawford AJ, Eggins SM, Zlobin SK. Primitive island arc and oceanic lavas from the Hunter Ridge-Hunter Fraction Zone. Evidence from glass, olivine, and spinal compositions. Mineral. Petrol. 1993;47:149–169.

Silver PG, Chan WW. Observations of body wave multipathing from broadband seismograms: evidence for lower mantle slab penetration beneath the Sea of Okhotsk. J. Geophys. Res. 1986;91:13787–13802.

Sisson TW, Layne GD. H_2O in basalt and basaltic andesite glass inclusions from four subduction-related volcanoes. Earth Planet. Sci. Lett. 1993;117:619–635.

Sleep NH. The double seismic zone in downgoing slabs and the viscosity of the mesosphere. J. Geophys. Res. 1979;84:4565–4571.

Sleep NH. Tapping of melt by veins and dikes. J. Geophys. Res. 1988;93:10255–10272.

Smith DC. Coesite in clinopyroxene in the Caledonides and its implications for geodynamics. Nature 1984;310:641–644.

Snoke JA, Sacks IS, James DE. Subduction beneath western South America: evidence from converted phases. Geophys. J. R. Astron. Soc. 1979;59:219–225.

Snoke JA, Sacks IS, Okada H. Determination of the subducting lithosphere boundary by use of converted phases. Bull. Seism. Soc. Am. 1977;67:1051–1060.

Song Y, Frey FA. Geochemistry of peridotite xenoliths in basalt from Hannuoba, Eastern China: Implications for subcontinental mantle heterogeneity. Geochim. Cosmochim. Acta 1989;53:97–113.

Sobolev AV, Shimizu N. Ultra-depleted primary melt included in an olivine from the mid-Atlantic ridge. Nature 1993;363:151–154.

Spakman W, Stein S, Hilst RDVd, Wortel R. Resolution experiments for NW pacific subduction zone tomography. Geophys. Res. Lett. 1989;16:1097–1100.

Spence DA, Turcotte DL. Buoyancy-driven magma fracture: a mechanism for ascent through the lithosphere and the emplacement of diamonds. J. Geophys. Res. 1990;95:5133–5139.

Spera FJ, Bergman SC. Carbon dioxide in igneous petrogenesis. I: aspects of dissolution of CO_2 in silicate liquids. Contrib. Mineral. Petrol. 1980;74:55–66.

Spiegelman M, McKenzie D. Simple 2-D models for melt extraction at mid-ocean ridges and island arcs. Earth Planet. Sci. Lett. 1987;83:137–152.

Staudigel H, Schreyer W. The upper thermal stability of clinochlore, $Mg_5Al(AlSi_3O_{10})(OH)_8$, at 10–35 kb P_{H_2O}. Contrib. Mineral. Petrol. 1977;61:187–198.

Stern CR, Frey FA, Futa K, Zartman RE, Peng Z, Kyser TK. Trace-element and Sr, Nd, Pb, and O isotopic composition of Pliocene and Quaternary alkali basalts of the Patagonian plateau lavas of southernmost South America. Contrib. Mineral. Petrol. 1990;104:294–308.

Stern RJ. A common mantle source for western Pacific island arc and "hot-spot" magmas—implications for layering in the upper mantle. Carnegie Inst. Wash. Yearb. 1981;81:455–462.

Stern TA. A backarc basin formed within continental lithosphere: the central volcanic region, New Zealand. Tectonophysics 1985;112:385–409.

Stern TA. Asymmetric backarc spreading, heat flux and structure associated with the central volcanic region of New Zealand. Earth Planet. Sci. Lett. 1987;85:265–276.

Stolz AJ, Varne R, Davies GR, Wheller GE, Foden JD. Magma source components in an arc-continent collision zone: the Flores-Lembata sector, Sunda Arc, Indonesia. Contrib. Mineral. Petrol. 1990;105:585–601.

Stolz AJ, Varne R, Wheller GE, Foden JD, Abbott MJ. The geochemistry and petrogenesis of K-rich alkaline volcanics from the Batu Tara volcano, eastern Sunda arc. Contrib. Mineral. Petrol. 1988;98:374–389.

Sudo A, Tatsumi Y. Phlogopite and K-amphibole in the upper mantle: implication for magma genesis in subduction zones. Geophys. Res. Lett. 1990;17:29–32.

Sugimura A. Zonal arrangement of some geophysical and petrological features in Japan and its environs. J. Fac. Sci. Univ. Tokyo Sect II 1960;12:133–153.

Sugimura A, Matsuda T, Chinzei K, Nakamura K. Quantitative distribution of late Cenozoic volcanic materials in Japan. Bull. Volcanol. 2 1963;26:125–140.

Sun S-s, McDonough WF. Chemical and isotopic systematics of oceanic basalts: implications for mantle composition and processes. In: Saunders AD, Norry

MJ, eds. Magmatism in the ocean basins. Geol. Soc. London Spec. Pub. 42, 1989:315–345.

Suzuki S, Sasatani T, Motoya Y. Double seismic zone beneath the middle of Hokkaido, Japan, in the southwestern side of the Kurile arc. Tectonophysics 1983;96:59–76.

Takada A. Experimental study on propagation of liquid-filled crack in gelatin: shape and velocity in hydrostatic stress condition. J. Geophys. Res. 1990;95:8471–8481.

Takagi A, Hasegawa A, Umino N. Seismic activity in the northeastern Japan arc. J. Phys. Earth 1977;25:S95–104.

Takahashi E. Melting relations of an alkali-olivine basalt to 30 kbar and their bearing on the origin of alkali basaltic magmas. Yearb. Carnegie Inst. Wash. 1980;79:271–276.

Takahashi E. Melting of a dry peridotite KLB-1 up to 14 GPa: implications on the origin of peridotitic upper mantle. J. Geophys. Res. 1986;91:9367–9382.

Takahashi E, Kushiro I. Melting of a dry peridotite at high pressures and basalt magma genesis. Am. Mineral. 1983;68:859–879.

Takahashi E, Uto K, Schilling J-G. Primary magma compositions and Mg/Fe ratios of their mantle residues along mid-Atlantic ridge 29°N to 73°N. Technical Report of ISEI Ser. A 1987;9:1–14.

Tapponnier P, Peltzer G, LeDain AY, Armijo R. Propagating extrusion tectonics in Asia: new insights from simple experiments with plasticine. Geology 1982;10:611–616.

Tarakanov RZ. On the possible role of seismofocal zones in formation and development of the island arc structure. Nauka, 1987:11–29.

Tarakanov RZ, Kim CU. Seismofocal zones and geodynamics of the Kuril-Japan region. In: Hilde TWC, Uyeda S, eds. Geodynamics of the western Pacific—Indonesian region. 11. Washington, DC: American Geophysical Union, Washington Geodyn, 1983:223–236.

Tatsumi Y. Melting experiments on a high-magnesian andesite. Earth Planet. Sci. Lett. 1981;54:357–365.

Tatsumi Y. Origin of high-magnesian andesites in the Setouchi volcanic belt, southwest Japan. II: melting experiments at high pressures. Earth Planet. Sci. Lett. 1982;60:305–317.

Tatsumi Y. High magnesian andesites in the Setouchi volcanic belt, southwest Japan and their possible relation to the evolutionary history of the Shikoku inter-arc basin. In: Hilde TWC, Uyeda S, eds. Geodynamics of the western Pacific—Indonesian region. 11. Washington, DC: American Geophysical Union, 1983:331–341.

Tatsumi Y. Migration of fluid phases and genesis of basalt magmas in subduction zones. J. Geophys. Res. 1989;94:4697–4707.

Tatsumi Y, Furukawa Y, Kogiso T, Yamanaka K, Yokoyama T, Fedotov SA. Unusual three volcanic chains in Kamchatka. Geophys. Res. Lett. 1994a.

Tatsumi Y, Furukawa Y, Yamashita S. Thermal and geochemical evolution of the mantle wedge in the NE Japan arc. I: contribution from experimental petrology. J. Geophys. Res. 1994b.

Tatsumi Y, Hamilton DL, Nesbitt RW. Chemical characteristics of fluid phase from a subducted lithosphere and origin of arc magmas: evidence from high-pressure experiments and natural rocks. J. Volcanol. Geotherm. Res. 1986;29:293–309.

Tatsumi Y, Ishizaka K. Existence of andesitic primary magma: an example from southwest Japan. Earth Planet. Sci. Lett. 1981;53:124–130.

Tatsumi Y, Ishizaka K. High magnesian andesite and basalt from Shodoshima island, southwest Japan, and their bearing on the genesis of calc-alkaline andesites. Lithos 1982a;15:161–172.

Tatsumi Y, Ishizaka K. Origin of high magnesian andesites in the Setouchi volcanic belt, southwest Japan. I: petrographical and chemical characteristics. Earth Planet. Sci. Lett. 1982b;60:293–304.

Tatsumi Y, Isoyama K. Transportation of beryllium with H_2O at high pressures and implication for magma genesis in subduction zones. Geophys. Res. Lett. 1988;15:180–183.

Tatsumi Y, Koyaguchi T. An absarokite from a phlogopite lherzolite source. Contrib. Mineral. Petrol. 1989;102:34–40.

Tatsumi Y, Maruyama S. Boninites and high-Mg andesites: tectonics and petrogenesis. In: Crawford AJ, ed. Boninites. London: Unwin Hyman, 1989:50–71.

Tatsumi Y, Murasaki M, Arsadi EM, Nohda S. Geochemistry of Quaternary lavas from NE Sulawesi: transfer of subduction components into the mantle wedge. Contrib. Mineral. Petrol. 1991;107:137–149.

Tatsumi Y, Murasaki M, Nohda S. Across-arc variation of lava chemistry in the Izu-bonin Arc: identification of subduction components. J. Volcanol. Geotherm. Res. 1992;49:179–190.

Tatsumi Y, Nakamura N. Composition of aqueous fluid from serpentinite in the subducted slab. Geochem. J. 1986;20:191–196.

Tatsumi Y, Nohda S, Ishizaka K. Secular variation of magma source compositions beneath NE Japan arc. Chem. Geol. 1988;68:309–316.

Tatsumi Y, Otofuji Y, Matsuda T, Nohda S. Opening of the Sea of Japan back-arc basin by asthenospheric injection. Tectonophysics 1989;166:317–329.

Tatsumi Y, Sakuyama M, Fukuyama H, Kushiro I. Generation of arc basalt magmas and thermal structure of the mantle wedge in subduction zones. J. Geophys. Res. 1983;88:5815–5825.

Tatsumi Y, Tsunakawa H. Cenozoic volcanism, stress gradient at back-arc opening in the North Island, New Zealand: origin of Tanpo-Rotorua Depression. The Island Arc, 1992:1:40–50.

Taylor B, Hayes DE. The tectonic evolution of the South China Basin. In: Hayes DE, ed. The tectonic and geologic evolution of Southeast Asian seas and islands. Geophys. Monogr., 23, Am. Geophys. Union, 1980:89–104.

Taylor B, Karner GD. On the evolution of marginal basins. Rev. Geophys. Space Phys. 1983;21:1727–1741.

Tera F, Brown L, Morris J, Sacks SI. Sediment incorporation in island-arc magmas: inference from ^{10}Be. Geochim. Cosmochim. Acta 1986;50:535–550.

Tera F, Brown J, Morris J, Sacks IS, Klein J, Middleton R. Sediment incorporation in island arc magmas: inference from Be-10. Geochim. Cosmochim. Acta 1985;60:636–660.

Thompson AB. Water in the Earth's upper mantle. Nature 1992;358:295–302.

Thompson RN. Primary basalts and magma genesis. Contrib. Mineral. Petrol. 1974;43:317–341.

Thorpe RS, Francis PW, Hammill M, Baker MCW. The Andes. In: Thorpe RS, ed. Andesites. New York: Wiley, 1982:187–205.

Tilley CE. Some aspects of magmatic evolution. Quant. J. Geol. Soc. London 1950;106:37–61.

Togashi S. Petrology of Miocene calc-alkaline rocks of northeastern Honshu, Japan. Sci. Rep. Tohoku Univ. Ser. III 1978;14:1–52.

Toksöz MN, Bird P. Formation and evolution of marginal basins and continental plateaus. In: Talwani M, Pitman IWC, eds. Island arcs, deep sea trenches and back-arc basins, Maurice Ewing Ser. 1. Washington, DC: American Geophysical Union, 1977:379–393.

Toksöz MN, Hsui AT. Numerical studies of back-arc convection and the formation of marginal basins. Tectonophysics 1978;50:177–196.

Toksöz MN, Minear JW, Julian BR. Temperature field and geophysical effects of a downgoing slab. J. Geophys. Res. 1971;76:1113–1138.

Tomita T. On the chemical compositions of the Cenozoic alkaline suite of the circum-Japan Sea region. J. Shanghai Sci. Inst, Sect II 1935;1:227–306.

Tomiya A. Volume of mantle diapir compatible with life span of a typical island-arc volcano. Bull. Volcanol. Soc. Japan. 1991;36:211–221.

Tse ST, Rice JR. Crustal earthquake instability in relation to the depth variation of frictional slip properties. J. Geophys. Res. 1986;91:9452–9472.

Tsunakawa H. Stress field during the rotation of southwest Japan. J. Geomagnet. Geoelectr. 1986;38:537–543.

Turner DL, Swanson SE, Wescott E. Continental rifting—a new tectonic model for geothermal exploration of the central Seward Peninsula, Alaska. Geotherm. Res. Council Trans. 1981;5:213–216.

Tyburczy JA, Waff H. Electrical conductivity of molten basalt and andesite to 25 kilobars pressure. J. Geophys. Res. 1983;88:2413–2430.

Ui T, Aramaki S. Relationship between chemical composition of Japanese island-arc volcanic rocks and gravimetric data. Tectonophysics 1978;45:249–259.

Umino N, Hasegawa A. On the two-layered structure of deep seismic plane in northeastern Japan arc. J. Seismol. Soc. Japan 1975;2-28:125–139.

Umino N, Hasegawa A. Three-dimensional Qs structure in the northeastern Japan arc. J. Seismol. Soc. Japan 1984;37:217–228.

Umino S, Kushiro I. Experimental studies on boninite petrogenesis. In: Crawford AJ, ed. Boninites. London: Unwin Hyman, 1989:89–111.

Utada H. A direct inversion method for 2-dimensional modeling in the geomagnetic induction problem. Doct. Thesis, Univ. Tokyo, 1988.

Utsu T. Anomalies in seismic wave velocity and attenuation associated with a deep earthquake zone. I. J. Fac. Sci. Hokkaido Univ., Ser. 7 1967;3:1–25.

Uyeda S. Subduction zones; facts, ideas and speculations. Oceanus 1979;22:52–62.

Uyeda S. Subduction zones: an introduction to comparative subductology. Tectonophysics 1982;81:133–159.

Uyeda S. Facts, ideas and open problems of trench-arc-backarc systems. In: Wezel FC, ed. The origin of arcs. Elsevier, 1986:435–460.

Uyeda S, Horai K. Terrestrial heat flow in Japan. J. Geophys. Res. 1964;69:2121–2141.

Uyeda S, Kanamori H. Backarc opening and the mode of subduction. J. Geophys. Res. 1979;84:1049–1061.

Van Bergen MJ, Vroon PZ, Varekamp JC, Poorter RPE. The origin of the potassic suite from Batu Tara volcano (East Sunda Arc, Indonesia). Lithos. 28, 1992:261–282.

Van Der Laan SR, Flower MFJ, Groos AF. Experimental evidence for the origin of boninites: near-liquidus phase relations to 7.5 kbar. In: Crawford AJ, ed. Boninites. London: Unwin Hyman, 1989:112–147.

Veith KF. The relationship of island arc seismicity to plate tectonics. Ph.D. thesis, Southern Methodist Univ. Dallas, TX. 1974:67.

Von Huene R, Scholl DW. Observations at convergent margins concerning sediment subduction, subduction erosion, and the growth of continental crust. Rev. Geophys. 1991;29:279–316.

Wadati K. On the activity of deep-focus earthquakes in the Japan Island and neighbourhoods. Geophys. Mag. 1955;8:305–325.

Waff HS, Rygh JT, Livelybrocks DW, Clingman WW. Results of a magnetotelluric traverse across western Oregon: crustal resistivity structure and the subduction of the Juan de Fuca plate. Earth Planet. Sci. Lett. 1988;87:313–324.

Walker D, Shibata T, Delong SE. Abyssal tholeiites from the oceanographer fracture zone. II: phase equilibria and mixing. Contrib. Mineral. Petrol. 1979;70:111–125.

Wang GF, Banno S. Non-stoichiometry of interlayer cations in micas from low- or middle-grade metamorphic rocks in the Ryoke and the Sanbagawa belts, Japan. Contrib. Mineral. Petrol. 1987;97:313–319.

Wang X, Liou JG, Mao HK. Coesite-bearing eclogite from the Dabie Mountains in central China. Geology 1989;17:1085–1088.

Wannamaker PE, Booker JR, Jones AG, et al. Resistivity cross section through the Juan de Fuca subduction system and its tectonic implication. J. Geophys. Res. 1989;94:14127–14144.

Watanabe T, Langseth MG, Anderson RN. Heat flow in back-arc basins of the western Pacific. In: Talwani M, Pitman IWC, eds. Island arcs, deep sea trenches, and back-arc basins, Maurice Ewing Ser. 1. Washington, DC: Am. Geophys. Union, 1977:137–167.

Weaver BL. The origin of ocean island basalt end-member compositions: trace element and isotopic constraints. Earth Planet. Sci. Lett. 1991;104:381–397.

Weaver CS, Baker GE. Geometry of the Juan de Fuca plate beneath Washington—evidence from seismicity and the 1949 South Puget Sound earthquake. Bull. Seism. Soc. Am. 1988;78:264–275.

Weaver SD, Smith IEM. New Zealand intraplate volcanism. In: Johnson RW, ed. Intraplate volcanism in eastern Australia and New Zealand. R. Soc. Austr. 1989:157–188.

Weidner DJ. A mineral physics test of a pyrolite mantle. Geophys. Res. Lett. 1985;12:417–420.

Wheller GE, Varne R, Foden JD, Abbott MJ. Geochemistry of Quaternary volcanics in the Sunda-Banda arc, Indonesia, and three component genesis of island arc basaltic magmas. J. Volcanol. Geotherm. Res. 1987;32:137–160.

White WM, Dupre B. Sediment subduction and magma genesis in the Lesser Antilles: isotopic and trace element constraints. J. Geophys. Res. 1986;91:5927–5941.

White WM, Patchett J. Hf-Nd Sr isotopes and incompatible element abundances in island arcs: implications for magma origins and crust-mantle evolution. Earth Planet. Sci. Lett. 1984;67:167–185.

Whitford DJ. Strontium isotopic studies of the volcanic rocks of the Sunda arc,

Indonesia, and their petrogenetic implications. Geochim. Cosmochim. Acta 1975;39:1287–1302.

Wilson M. Igneous Petrogenesis. London: Unwin Hyman, 1989.

Woodhead J, Eggins S, Gamble JA. High field strength and transition element systematics in island arc and back-arc basin basalts: evidence for multi-phase melt extraction and a depleted mantle wedge. Earth Planet. Sci. Lett. 1993;114:491–504.

Woodhead JD, Fraser DG. Pb, Sr and ^{10}Be isotopic studies of volcanic rocks from the northern Mariana islands: implications for magma genesis and crustal recycling in the western Pacific. Geochim. Cosmochim. Acta 1985;49:1925–1930.

Woodhead JD, Johnson RW. Isotopic and trace element profiles across the New Britain island arc, Papua, New Guinea. Contrib. Mineral. Petrol. 1993;113:479–491.

Wunder B, Schreyer W. Metastability of the 10-Å phase in the system MgO-SiO_2-H_2O (MSH). What about hydrous MSH phases in subduction zones? J. Petrol. 1992;33:877–889.

Wyllie PJ. Crustal anatexis: an experimental review. Tectonophysics 1977;43:41–71.

Wyllie PJ. Magma genesis, plate tectonics, and chemical differentiation of the earth. Rev. Geophys. 1988;26:370–404.

Wyllie PJ, Sekine T. The formation of mantle phlogopite in subduction zone hybridization. Contrib Mineral Petrol 1982;79:375–380.

Xie GH, Wang WJ, Basu AR, Tatsumoto M. Petrochemistry and Sr, Nd, Pb, isotopic geochemistry of Cenozoic volcanic rocks, Changbaishan area, northeast China. Acta Petrol. Sinica 1988;4:1–13.

Yamamoto K, Akimoto S-I. The system MgO-SiO_2-H_2O at high pressures and temperatures—stability field of hydroxyl-chondrodite, hydroxyl-clinohumite and 10 Å phase. Am. J. Sci. 1977;277:288–312.

Yamamoto M. Picritic primary magma and its source mantle for Oshima-Oshima and back-arc side volcanoes, northeast Japan arc. Contrib. Mineral. Petrol. 1988;99:352–359.

Yamano M, Uyeda S, Honda S. Nankai Trough: a hot trench? Mar. Geophys. Res. 1984;6:187–203.

Yamaoka K, Fukao Y. Why do island arcs form cusps at their junctions? Geology 1987;15:34–36.

Yamaoka K, Fukao Y, Kumazawa M. Spherical shell tectonics: effects of sphericity and inextensibility on the geometry of the descending lithosphere. Rev. Geophys. 1986;24:27–53.

Yamashita S, Fujii T. Experimental petrology of basement basaltic rocks from Sites 794 and 797, Japan Sea. In: Tamaki K, Ingle J, Suyehiro K, et al. eds. Proc. ODP, Sci. Res. 127/128. College Station, TX: Ocean Drilling Program, 1992:891–898.

Yamashita S, Tatsumi Y. Thermal and geochemical evolution of mantle wedge in NE Japan arc. II: contribution from geochemistry. J. Geophys. Res. 1994;99:22285–22293.

Yoder HS, Stewart DB, Smith JR. Ternary feldspars. Carnegie Inst. Wash. Yearb. 1957;56:206–214.

Yoder HS, Tilley CE. Origin of basalt magmas: an experimental study of natural and synthetic rock systems. J. Petrol. 1962;3:342–532.

Yoshii T. Proposal of the "aseismic front." J. Seismol. Soc. Japan 1975;2-28:365–367.

Yoshii T, Asano S. Time-term analyses at explosion seismic data. J. Phys. Earth 1972;20:47–57.

You C-F, Spivack AJ, Smith JH, Gleskes JM. Mobilization of boron in convergent margins: implications for the boron geochemical cycle. Geology 1993;21:207–210.

Yuen DA, Fleitout L, Schubert G, Froidevaux C. Shear deformation zones along major transform faults and subducting slabs. Geophys. J. R. Astron. Soc. 1978;54:93–119.

Yuen DA, Schubert G. On the stability field of frictionally heated shear flows in the asthenosphere. Geophys. J. R. Astron. Soc. 1979;57:189–207.

Zhao D. A tomographic study of seismic velocity structure in the Japan Islands. Ph.D. thesis, Tohoku Univ. 1991:301.

Zhao D, Horiuchi S, Hasegawa A. 3-D seismic velocity structure of the crust and the uppermost mantle in the northeastern Japan arc. Tectonophysics 1990;181:135–149.

Zhao W, Yuen DA, Honda S. Multiple phase transitions and the style of mantle convection. Phys. Earth Planet. Inter. 1992;72:185–210.

Zhou HW. Mapping of P-wave slab anomalies beneath the Tonga, Kermadec and New Hebrides arcs. Phys. Earth Planet. Inter. 1990;61:199–229.

Zhou HW, Clayton RW. P and S travel time inversions for subducting slab under the island arcs of the northwest Pacific. J. Geophys. Res. 1990;95:6829–6851.

Zhuravlev DZ, Tsverkov AA, Gladkov NG, Chernyshev IV. $^{143}Nd/^{144}Nd$ and $^{87}Sr/^{86}Sr$ ratios in recent magmatic rocks of the Kurile island arc. Chem. Geol. 1987;66:224–243.

Ziagos J, Blackwell D, Mooser F. Heat flow in southern Mexico and the thermal effects of subduction. J. Geophys. Res. 1985;90:5410–5420.

Zoback ML, Zoback M. State of stress in the conterminous United States. J. Geophys. Res. 1980;85:6113–6156.

Index

N

O

P

R

S

T